Roland H. Tarkhanyan and
Nikolaos K. Uzunoglu

Radiowaves and Polaritons
in Anisotropic Media

Related Titles

Yeh, P.

Optical Waves in Layered Media

406 pages, Softcover
2005
ISBN 0-471-73192-7

White, J. F.

High Frequency Techniques
An Introduction to RF and Microwave Engineering

502 pages, Hardcover
2004
ISBN 0-471-45591-1

Bonnet, M.

Boundary Integral Equation Methods for Solids and Fluids

412 pages, Hardcover
1999
ISBN 0-471-97184-7

Ishimaru, A.

Wave Propagation and Scattering in Random Media

600 pages, Softcover
1999
ISBN 0-7803-4717-X

Chew, W.C.

Waves and Fields in Inhomogenous Media

632 pages, Softcover
1999
ISBN 0-7803-4749-8

Jones, D.S.J.

Methods in Electromagnetic Wave Propagation
2nd Edition

672 pages, Hardcover
1994
ISBN 0-7803-1155-8

Roland H. Tarkhanyan and Nikolaos K. Uzunoglu

Radiowaves and Polaritons in Anisotropic Media

WILEY-VCH

WILEY-VCH Verlag GmbH & Co. KGaA

Editor

Roland H. Tarkhanyan
National Academy of Sciences of Armenia
Institute of Radiophysics and Electronics
Yerevan, Republic of Armenia
rolandtarkhanyan@yahoo.com

Nikolaos K. Uzunoglu
National Technical University of Athens
School of Electrical and Computer Engineering
Division of Information Transmission Systems and Materials Technology
Mircrowave and Fiber Optics Laboratory
Athens, Greece
nuzu@cc.ece.ntua.gr

Cover
Dr. Bäsemann/Naturfoto-Online

Library of Congress Card No.: applied for

British Library Cataloguing-in-Publication Data
A catalogue record for this book is available from the British Library.

Bibliographic information published by Die Deutsche Bibliothek
Die Deutsche Bibliothek lists this publication in the Deutsche Nationalbibliografie; detailed bibliographic data is available in the Internet at <http://dnb.ddb.de>.

Printed in the Federal Republic of Germany.
Printed on acid-free paper.

Typesetting Kühn & Weyh, Satz und Medien, Freiburg
Printing Strauss GmbH, Mörlenbach
Binding Litges & Dopf Buchbinderei GmbH, Heppenheim

ISBN-13: 978-3-527-40615-9
ISBN-10: 3-527-40615-8

Contents

Radiowaves and Polaritons in Anisotropic Media. Roland H. Tarkhanyan and Nikolaos K. Uzunoglu

ISBN: 3-527-40615-8

Abstract

A theory of propagation and excitation of volume and surface electromagnetic waves in anisotropic polar and nonpolar conducting crystals is presented. Effects of external magnetic and strong electric fields are considered. The spectra of bulk as well as surface phonon-plasmon polaritons in uniaxial semiconductors are investigated. Effects caused by the interface magnon-plasmon polaritons in layered structures semiconductor/magnetic insulator and superlattices are presented. Electromagnetic instabilities leading to the generation and amplification of UHF radio waves and guided waves in semiconductor heterostructures with hot two-dimensional charge carriers are studied.

A number of original and interesting phenomena are considered which have not been presented in the monographical literature formerly. In particular, the influence of the anisotropy of parameters describing the crystal lattice and free charge carriers on the high-frequency electromagnetic instabilities in semiconductors with positive differential conductivity is investigated. Radiation of a dipole source in the presence of a grounded gyromagnetic slab is studied.

This book is written for specialists in solid-state and semiconductor physics, as well as for electrical and materials engineers, who want to gain a fundamental understanding of the optical properties of anisotropic materials. In addition, it can be useful for students learning applied electrodynamics, crystallooptics and microelectronics.

Radiowaves and Polaritons in Anisotropic Media. Roland H. Tarkhanyan and Nikolaos K. Uzunoglu

ISBN: 3-527-40615-8

Preface

The present book on propagation and excitation of electromagnetic waves in anisotropic conducting media is, in many aspects, different from other books on solid-state physics. First of all, our major interest lies in showing how the anisotropy of parameters describing the crystal lattice as well as the free charge carriers (electrons or holes in semiconductors) cause new physical phenomena which are absent in isotropic media. Second, we give a more detailed description of the interaction of electromagnetic waves (radio waves, millimeter and submillimeter waves, far-infrared radiation) with different vibrational subsystems in anisotropic materials: electronic plasma longitudinal vibrations (in semiconductors), optical phonons (in polar crystals), optical magnons (in antiferromagnetic insulators) and spin waves (in ferrites), as well as the possible combinations of these vibrations (that is, bulk phonon-plasmon and interface magnon-plasmon polaritons). Third, this monograph is not an encyclopedia. The selection of topics is restricted to material which is considered to be essential and which is partly due to our personal inclinations. Fourth, this book is distinctly divided into four self-contained parts which may be read independently. Parts 1-3 are written by R.H. Tarkhanyan and Part 4 by N.K. Uzunoglu.

Part 1 includes more details of volume electromagnetic waves in uniaxial polar crystals with electronic plasma, as well as the influence of an external magnetic field on the propagation and reflection of the waves. Part 2 introduces the reader to the fundamentals of the surface and interface electromagnetic wave theory, in different semiconductor structures. In particular, propagation of the surface waves in a ferrite/semiconductor superlattice at quantum Hall effect conditions is considered. Basic concepts of the electromagnetic instabilities in uniaxial semiconductors with hot carriers are introduced in Part 3. It contains some original results on generation and amplification of the waves and a number of interesting phenomena which have not been presented in monographs yet. For example, instability of the waves in the absence of a falling region in the current-voltage characteristic is considered. In Part 4, radiation of a dipole source in a grounded gyromagnetic slab is considered, using the Green function method.

Finally, this book is written for specialists and postgraduate students in solid-state and semiconductor physics, as well as for engineers, particularly electrical and materials engineers, who want to gain a fundamental understanding of the

Radiowaves and Polaritons in Anisotropic Media. Roland H. Tarkhanyan and Nikolaos K. Uzunoglu

ISBN: 3-527-40615-8

optical properties of anisotropic semiconductor materials and devices. Thus, we tried to bridge the gap between physics and engineering. That is why our book stresses concepts rather than mathematical formalism, which should make the presentation relatively easy to understand. Nevertheless, it is assumed that the reader has taken courses in solid-state and semiconductor physics, classical electromagnetics and crystallooptics. In addition, the reader should be familiar with the necessary mathematics in vector and tensor calculations.

Roland H. Tarkhanyan
Nikolaos K. Uzunoglu

List of Contributors

Roland H. Tarkhanyan
Institute of Radiophysics and
Electronics of the Armenian National
Academy of Sciences Ashtarack-2
378410
Armenia

Nikolaos K. Uzunoglu
National Technical University of Athens
9 Iroon Politechniou Str.
15780 Zografos
Athens
Greece

R.H. Tarkhanyan received a Ph.D. degree in physics in 1966 from A.F. Ioffe Physico-Technical Institute (St. Petersburg, Russia) and a Dr. Habil. degree in physics and mathematics in 1980 from the High Examining Board in the Ministerial Council of the USSR, Moscow. Since 1966 he has been with the Institute of Radiophysics and Electronics of the Armenian National Academy of Sciences, where he now holds the position of Leading Scientist. He also was Professor at Yerevan State University and Visiting Researcher at the Technical University of Athens and at the Greek National Research Center "Demokritos". He is the author of 91 publications in refereed international journals on theoretical solid-state and semiconductor physics.

N.K. Uzunoglu received his M.Sc. and Ph.D. degrees in 1974 and 1976 from the University of Essex, UK. Since 1987 he has been Professor at the National Technical University of Athens, Department of Electrical Engineering, where he was elected Chairman twice. His research interests include electromagnetic scattering radiation phenomena, fiber optics telecommunications and high-speed circuits. In 1981 he obtained the International G. Marconi award in telecommunications.

Radiowaves and Polaritons in Anisotropic Media. Roland H. Tarkhanyan and Nikolaos K. Uzunoglu

ISBN: 3-527-40615-8

He has over 200 publications in journals and he has published three books in Greek on microwaves, fiber optics telecommunications and radar systems.

Part 1
Volume Electromagnetic Waves in Anisotropic Crystals with Electronic Plasma

Roland H. Tarkhanyan

Radiowaves and Polaritons in Anisotropic Media. Roland H. Tarkhanyan and Nikolaos K. Uzunoglu

ISBN: 3-527-40615-8

Introduction

Propagation of electromagnetic waves in crystalline solids is the basis for understanding their optical properties in different ranges of frequencies and wavelengths. A large number of books are now available, concerning specific properties of the waves in solids, and in most of them crystals of cubic structure with isotropic optical behavior are considered. However, most semiconductors and ionic insulators possess anisotropic crystal structures. It is well known that 19 from 32 possible point groups for crystals are characterized by the presence of the axis of rotation of the third, fourth or sixth order [1]. All these crystals relate to the group of optically uniaxial crystals, the optical properties of which are described by the permittivity tensor with two different principal eigenvalues along and across the crystal axis of symmetry. If a uniaxial crystal has a center of symmetry, that is, a point through which the crystal may be inverted and left invariant, the opposite directions along the axis of symmetry are equivalent.

A great number of general properties of electromagnetic waves in anisotropic nonconducting media have been discussed in Refs. [2–6], which also hold for uniaxially stressed silicon and germanium, Te, Se, $CdAs_2$, InSe, α-SiC(6H), CdS and other uniaxial semiconductors. However, the presence of the free carriers and phonons and their interaction with the electromagnetic waves lead to new specific properties of the waves, which are absent in anisotropic insulators [2–6] as well as in metals [7–9].

Part 1 of this book develops the fundamentals of a phenomenological theory of bulk electromagnetic waves in both nonpolar and polar (ionic) uniaxial crystals with electronic plasma. It is assumed that isoenergetic surfaces for the conduction electrons are ellipsoids of rotation. A great number of new physical effects caused by the anisotropy of the parameters describing both the crystal lattice and the free charge carriers are considered.

For simplicity, we will neglect the interband transitions and the space dispersion, that is, the wave-number dependence in both the conductivity and the lattice dielectric permittivity tensor. Thus, we will restrict ourselves to the cases of the long-wavelength limit for the lattice vibrations and "cold" electronic plasma. It must be mentioned that, as a result of the appearance of new waves due to the spatial dispersion, the ordinary continuity conditions for the field components on the crystal surface are found to be insufficient for the joining up of the fields and

Radiowaves and Polaritons in Anisotropic Media. Roland H. Tarkhanyan and Nikolaos K. Uzunoglu

ISBN: 3-527-40615-8

the calculation of the reflection coefficient, and additional boundary conditions must be introduced [10–20]. However, a detailed discussion of matters arising from the spatial dispersion lies beyond the scope of the present book.

In the chapters to come, we will first consider the influence of the anisotropy on the propagation of "pure" electromagnetic, longitudinal plasma and long-wavelength optical vibrations and then the spectrum of the bulk polaritons in uniaxial polar semiconductors [21, 22]. Similar problems in the isotropic case have been considered in Refs. [23–27]. Radio waves and polaritons in uniaxial semiconductors in the presence of an external static magnetic field are discussed in Chapter 3 for both the Faraday and the Voigt configurations [28, 29]. We conclude Part 1 by consideration of the wave reflection and refraction problem on the surface of uniaxial semiconductors, in the absence as well as in the presence of a static magnetic field.

1
Influence of the Anisotropy on the Spectrum and Propagation of Electromagnetic, Plasma and Lattice Optical Vibrations

1.1
Maxwell's Equations and High-Frequency Conductivity Tensor of an Anisotropic Semiconductor

Consider an anisotropic nonpolar and nonmagnetic semiconductor with electron or hole plasma. For brevity of the following presentation we consider only one plasma component, for instance the electron component. Let us assume that iso-energetic surfaces for the free charge carriers are ellipsoids of rotation. Then, we can write the energy of the conduction electrons in the form

$$W(\boldsymbol{p}) = \frac{\mu_{\alpha\beta} p_\alpha p_\beta}{2m} = \frac{1}{2m}\left(\mu_\perp p_\perp^2 + \mu_{||} p_{||}^2\right) \tag{1.1}$$

where m is the free electron mass, $\mu_{\alpha\beta}$ is the dimensionless reciprocal effective mass tensor with two different eigenvalues $\mu_{||}$ and $\mu_\perp$ along and across the crystal axis, $p_{||}$ and $p_\perp$ are the corresponding components of the quasimomentum and α, $\beta = x, y, z$. We are using the convention that a sum is implied over repeated indices.

The electromagnetic properties of the crystal are described by Maxwell's equations

$$\text{rot}\, \boldsymbol{H} = \boldsymbol{J} + \frac{\partial}{\partial t}\boldsymbol{D}_\text{L} \tag{1.2}$$

$$\text{rot}\, \boldsymbol{E} = -\frac{\partial}{\partial t}\boldsymbol{B}$$

where $\boldsymbol{E}$ and $\boldsymbol{H}$ are the electric and magnetic field intensities,

$$\boldsymbol{D}_\text{L} = \varepsilon_0 \varepsilon_\text{L} \boldsymbol{E} \tag{1.4}$$

$$\boldsymbol{B} = \mu_0 \boldsymbol{H} \tag{1.5}$$

Radiowaves and Polaritons in Anisotropic Media. Roland H. Tarkhanyan and Nikolaos K. Uzunoglu

ISBN: 3-527-40615-8

are the electric and magnetic flux densities, $\boldsymbol{\varepsilon}_L$ is the dielectric permittivity tensor of the crystal lattice with constant eigenvalues $\varepsilon_{||}^L$ and $\varepsilon_{\perp}^L$ along and across the crystal axis, respectively, ε_0 and μ_0 are the permittivity and permeability of free space and

$$J(\boldsymbol{r}, t) = e\int \boldsymbol{v} f(\boldsymbol{r}, \boldsymbol{p}, t) d\boldsymbol{p} \tag{1.6}$$

is the electric current density. The function $f(\boldsymbol{r}, \boldsymbol{p}, t)$ under the integral in Eq. (1.6) is a small perturbation of the carrier's momentum distribution function $f_0(\boldsymbol{p})$ in the equilibrium state,

$$d\boldsymbol{p} = \frac{2}{(2\pi\hbar)^3} dp_x dp_y dp_z \tag{1.7}$$

$$\boldsymbol{v} = dW(\boldsymbol{p})/d\boldsymbol{p} = \boldsymbol{\mu p}/m \tag{1.8}$$

is the velocity vector and "e" denotes the charge of the free carriers. Usually the function $f(\boldsymbol{r}, \boldsymbol{p}, t)$ is obtained by solving the linearized kinetic equation [30]

$$\partial f/\partial t + \boldsymbol{v}\nabla f + e(\boldsymbol{E} + [\boldsymbol{vB}])\partial f_0/\partial \boldsymbol{p} + f/\tau = 0 \tag{1.9}$$

where $\tau^{-1} = \nu$ is the scattering frequency of the charge carriers. In frame of linear electrodynamics, using the Fourier transformation, the self-consistent solutions of the system of equations (1.2)–(1.9) can be sought as $\exp(-i\omega t + i\boldsymbol{kr})$. Then, taking into account the relation

$$\partial f_0/\partial \boldsymbol{p} = \boldsymbol{v}\partial f_0/\partial w \tag{1.10}$$

for the Fourier components of the quantities under consideration, one can obtain

$$f(\omega, \boldsymbol{k}, \boldsymbol{p}) = -ie\boldsymbol{E}(\omega, \boldsymbol{k})(\omega + i\nu - \boldsymbol{kv})^{-1}\partial f_0/\partial \boldsymbol{p} \tag{1.11}$$

$$\boldsymbol{J}(\omega, \boldsymbol{k}) = \boldsymbol{\sigma}(\omega, \boldsymbol{k})\boldsymbol{E}(\omega, \boldsymbol{k}) \tag{1.12}$$

where the conductivity tensor is

$$\sigma_{\alpha\beta}(\omega, \boldsymbol{k}) = -ie^2\int v_\alpha v_\beta(\omega + i\nu - \boldsymbol{kv})^{-1}\frac{\partial f_0}{\partial w}d\boldsymbol{p} \tag{1.13}$$

We shall be interested in high-frequency waves, for which the conditions

$$\omega \gg \nu \tag{1.14}$$

and $\quad \omega \gg \boldsymbol{kv} \quad$ (1.15)

are fulfilled. In this case the dissipation of the waves due to the scattering processes as well as the spatial dispersion in Eq. (1.13) can be neglected, and the high-frequency conductivity tensor can be written in the form

$$\sigma_{\alpha\beta}(\omega) = -\mathrm{i}\frac{e^2}{\omega}\int v_\alpha v_\beta \frac{\partial f_0}{\partial w}\mathrm{d}\boldsymbol{p} \tag{1.16}$$

When the electron gas is nondegenerate (see next section), this tensor has a simple form

$$\sigma_{\alpha\beta}(\omega) = \frac{\mathrm{i}Ne^2}{m\omega}\mu_{\alpha\beta} \tag{1.16a}$$

where N is the density of the free carriers given by Eq. (1.21), below.

1.2 Complex Dielectric Permittivity Tensor

Let us introduce now the total electric flux density

$$\boldsymbol{D}(\boldsymbol{r},t) = \boldsymbol{D}_\mathrm{L}(\boldsymbol{r},t) + \int_{-\infty}^{t} \boldsymbol{j}(\boldsymbol{r},t')\mathrm{d}t' \tag{1.17}$$

Then Eq. (1.2) can be written in the form

$$\mathrm{rot}\ \boldsymbol{H} = \partial\boldsymbol{D}/\partial t \tag{1.2a}$$

For plane waves, using Eqs. (1.4), (1.12) and (1.17), one can obtain

$$\boldsymbol{D}(\omega,\boldsymbol{k}) = \varepsilon_0\boldsymbol{\varepsilon}(\omega,\boldsymbol{k})\boldsymbol{E}(\omega,\boldsymbol{k}) \tag{1.18}$$

where $$\boldsymbol{\varepsilon}(\omega,\boldsymbol{k}) = \varepsilon_\mathrm{L} + \frac{\mathrm{i}}{\varepsilon_0\omega}\boldsymbol{\sigma}(\omega,\boldsymbol{k}) \tag{1.19}$$

is the complex dielectric permittivity tensor, which connects the Fourier components of the electric field intensity and the electric flux density.

In the case of Maxwellian statistics (nondegenerate electrons), for the equilibrium distribution function we have [30]

$$f_0(\boldsymbol{p}) = \mu_\perp\mu_{||}{}^{1/2}N(2\pi mT)^{-3/2}\exp[-w(\boldsymbol{p})/T] \tag{1.20}$$

where T is the temperature in energetic units and

$$N = \int f_0(\boldsymbol{p})\mathrm{d}\boldsymbol{p} \tag{1.21}$$

is the density of the conduction electrons.

Using Eqs. (1.13) and (1.20), for the components of the permittivity tensor (1.19) in the collisionless case we obtain, in the coordinate system of the principal axis of the tensor $\mu_{\alpha\beta}$, the following expression:

$$\varepsilon_{\alpha\beta} = \varepsilon_a^{\rm L}\delta_{\alpha\beta} + \frac{\omega_0^2}{\omega^2}\sqrt{\mu_\alpha\mu_\beta}\{zF(z)\delta_{\alpha\beta} + \frac{\kappa_\alpha\kappa_\beta}{\kappa^2}[2z^2 + (2z^2 - 1)zF(z)]\} \tag{1.22}$$

where

$$\omega_0^2 = \frac{Ne^2}{m\varepsilon_0} \tag{1.23}$$

$$z = \frac{\omega}{\kappa v_T},\ v_T = \sqrt{\frac{2T}{m}},\quad \boldsymbol{\kappa} = \{k_x\sqrt{\mu_\perp}, k_y\sqrt{\mu_\perp}, k_z\sqrt{\mu_{||}}\} \tag{1.24}$$

$$F(z) = -\frac{1}{\sqrt{\pi}}\int_{-\infty}^{\infty}\frac{{\rm e}^{-t^2}}{z-t}{\rm d}t = i\sqrt{\pi}{\rm e}^{-z^2} - 2{\rm e}^{-z^2}\int_0^z {\rm e}^{t^2}{\rm d}t \tag{1.25}$$

is the plasma dispersion function [31] and $\delta_{\alpha\beta}$ is the Kronecker δ-symbol. The imaginary part in Eq. (1.25) represents the collisionless Landau dissipation of the waves by resonant particles. For a weak spatial dispersion, when the condition

$$z \gg 1 \tag{1.26}$$

is fulfilled, the Landau dissipation can be neglected and one can use the asymptotic expression

$$-zF(z) \approx 1 + \frac{1}{2z^2} + \frac{3}{4z^4} + \ldots \tag{1.27}$$

Then, for the components of the complex dielectric permittivity tensor given in Eq. (1.22) we obtain, with an accuracy of the terms proportional to z^{-2},

$$\varepsilon_{\alpha\beta} = [\varepsilon_a^{\rm L} - \mu_a\frac{\omega_0^2}{\omega^2}(1 + \frac{\mu_a K_d^2 V_T^2}{2\omega^2})]\delta_{\alpha\beta} - \left(\frac{\omega_0}{\omega}\right)^2\frac{\mu_a k_a}{\omega^2}\mu_\beta k_\beta v_T^2 \tag{1.28}$$

In the limiting case $kv_T/\omega \to 0$ the spatial dispersion can be neglected. Then, the nondiagonal components of the tensor vanish and $\varepsilon_{\alpha\beta}$ can be written in the form

$$\varepsilon = \begin{pmatrix} \varepsilon_\perp(\omega) & 0 & 0 \\ 0 & \varepsilon_\perp(\omega) & 0 \\ 0 & 0 & \varepsilon_{||}(\omega) \end{pmatrix} \tag{1.29}$$

where

$$\varepsilon_{\perp}(\omega) = \varepsilon_{\perp}^{L}\left(1 - \frac{\omega_{0\perp}^2}{\omega^2}\right), \ \omega_{0\perp}^2 = \frac{\mu_{\perp}}{\varepsilon_{\perp}^{L}}\omega_0^2 \quad (1.30)$$

$$\varepsilon_{\parallel}(\omega) = \varepsilon_{\parallel}^{L}\left(1 - \frac{\omega_{0\parallel}^2}{\omega^2}\right), \ \omega_{0\parallel}^2 = \frac{\mu_{\parallel}}{\varepsilon_{\parallel}^{L}}\omega_0^2 \quad (1.31)$$

and ω_0^2 is given in Eq. (1.23).

1.3 Dispersion Relations for Electromagnetic Waves. Regions of Propagation. Resonances and Cut-Off Frequencies

Let us now find dispersion relations which determine the functional dependence between the frequency ω and the wave vector $\boldsymbol{k}$ of the electromagnetic waves in uniaxial crystals with electronic plasma. For the uniform plane waves for which $\boldsymbol{E}$, $\boldsymbol{H}$ ~exp[i$\boldsymbol{kr}$ – iωt], the Maxwell equations become

$$[\boldsymbol{kH}] = -\omega\, \boldsymbol{D} \quad (1.32)$$

$$[\boldsymbol{kE}] = \omega\, \boldsymbol{B} \quad (1.33)$$

Using the constitutive relations (1.5) and (1.18) and eliminating the magnetic field

$$\boldsymbol{H} = (\omega\mu_0)^{-1}[\boldsymbol{kE}] \quad (1.33a)$$

from Eq. (1.32) we obtain

$$k^2\boldsymbol{E} - \boldsymbol{k}(\boldsymbol{kE}) - \frac{\omega^2}{c^2}\varepsilon\boldsymbol{E} = 0 \quad (1.34)$$

where $c = (\varepsilon_0\mu_0)^{-1/2}$ is the light velocity in free space. It is convenient to write Eq. (1.34) in the form

$$(n^2\delta_{\alpha\beta} - n_\alpha n_\beta - \varepsilon_{\alpha\beta})E_\beta = 0 \quad (1.35)$$

where

$$\boldsymbol{n} = n\boldsymbol{s} \quad n = ck/\omega \quad (1.36)$$

is the refractive index and $\boldsymbol{s} = \boldsymbol{k}/k$ is the unit vector in the direction of the wave propagation.

The system of equations (1.35) has a nontrivial solution only if the determinant of the coefficient matrix vanishes:

$$\det[\varepsilon_{\alpha\beta} + n^2(s_\alpha s_\beta - \delta_{\alpha\beta})] = 0 \tag{1.37}$$

Let us choose the z-axis of the coordinate system along the crystal axis and assume that the wave is propagating in the xz-plane, so that $s_y = 0$. Then, using Eq. (1.29) for the tensor $\varepsilon_{\alpha\beta}$ we can write Eq. (1.37) in the form

$$(n^2 - \varepsilon_\perp)\left(n^2 - \frac{\varepsilon_\perp \varepsilon_{||}}{\varepsilon_\perp \sin^2\theta + \varepsilon_{||}\cos^2\theta}\right) = 0 \tag{1.38}$$

where θ is the angle between the crystal axis and the wave vector $\boldsymbol{k}$. It follows that along a given direction in a uniaxial crystal there can propagate two different and independent electromagnetic waves: an ordinary wave which is described by the dispersion relation

$$n^2 = \varepsilon_\perp(\omega) \tag{1.39}$$

and an extraordinary wave for which

$$n^2 = \frac{\varepsilon_\perp(\omega)\varepsilon_{||}(\omega)}{\varepsilon_\perp(\omega)\sin^2\theta + \varepsilon_{||}(\omega)\cos^2\theta}. \tag{1.40}$$

Let us now examine their properties separately. Using Eq. (1.34), it is not difficult to see that the ordinary wave is a transverse (or TE-polarized) one: $\boldsymbol{E} = \{0, E_y, 0\}$. The refractive index, as in the case of an isotropic medium, has no dependence on the direction of propagation. Using Eq. (1.30) for $\varepsilon_\perp(\omega)$, from Eq. (1.39) we obtain

$$\omega^2 = \omega_{0\perp}^2 + c^2k^2/\varepsilon_\perp^{\mathrm{L}}. \tag{1.41}$$

It follows that the ordinary wave can only propagate in the frequency region

$$\omega > \omega_{0\perp} \tag{1.42}$$

At $\omega = \omega_{0\perp}$ the wave has a cut-off: $k = 0$. At the cut-off frequency both the magnetic field in the wave and the total electric flux density $\boldsymbol{D} = \varepsilon_\perp(\omega)\boldsymbol{E}$ vanish because the displacement current $\partial \boldsymbol{D}_{\mathrm{L}}/\partial t$ in Eq. (1.2) is completely compensated by the conductivity current $\boldsymbol{J}$.

In a particular case of a nonconducting material, the cut-off frequency vanishes, and then Eq. (1.41) gives

$$\omega = \frac{ck}{\sqrt{\varepsilon_\perp^{\mathrm{L}}}} \tag{1.41a}$$

As to the extraordinary wave, it has TM-type polarization: $\boldsymbol{H} = \{0, H_y, 0\}$. The electric field vector $\boldsymbol{E}$ lies in the xz-plane and makes an angle φ_E with the crystal axis, where

$$\tan\varphi_E = -\frac{\varepsilon_{||}(\omega)}{\varepsilon_{\perp}(\omega)}\cot\theta \tag{1.43}$$

Thus, both the direction of the vector $\boldsymbol{E}$ and the refractive index of the extraordinary wave are essentially dependent on the direction of propagation. When the wave propagates along ($\theta = 0$) or across ($\theta = \pi/2$) the crystal axis, it splits into a longitudinal ($\boldsymbol{E}||\boldsymbol{k}$) and a transversal ($\boldsymbol{E}\perp\boldsymbol{k}$) wave with dispersion relations given respectively by

$$\theta = 0: \quad \varepsilon_{||}(\omega) = 0, \quad n^2 = \varepsilon_{\perp}(\omega). \tag{1.44a}$$

$$\theta = \pi/2: \quad \varepsilon_{\perp}(\omega) = 0, \quad n^2 = \varepsilon_{||}(\omega). \tag{1.44b}$$

The longitudinal waves possess characteristic plasma frequencies given in Eqs. (1.30) and (1.31).

For $\theta \neq 0, \pi/2$ the extraordinary wave propagates ($k^2 > 0$) only in the following two frequency regions:

$$\min(\omega_{0\perp}, \omega_{0||}) < \omega < \omega_{\mathrm{r}} \tag{1.45}$$

$$\omega > \max(\omega_{0\perp}, \omega_{0||}) \tag{1.46}$$

where

$$\omega_{\mathrm{r}} = \omega_0 \left(\frac{\mu_{\perp}\sin^2\theta + \mu_{||}\cos^2\theta}{\varepsilon_{\perp}^{\mathrm{L}}\sin^2\theta + \varepsilon_{||}^{\mathrm{L}}\cos^2\theta}\right)^{1/2} \tag{1.47}$$

is the resonance frequency ($k \to \infty$). The wave has two different cut-off frequencies: $\omega = \omega_{0||}$ and $\omega = \omega_{0\perp}$. The resonance frequency, in contrast to cut-off frequencies, depends essentially on the direction of propagation.

Thus, even in the above-considered simplest case, the anisotropy of the crystal leads to a noticeable effect, namely to the appearance of two transparent frequency regions (contrary to one region in the case of isotropic media), that are separated by a nontransparent region

$$\omega_{\mathrm{r}} < \omega < \max(\omega_{0\perp}, \omega_{0||}) \tag{1.48}$$

with the width depending on the direction of propagation. We have to note that the splitting of the transparent region into two occurs only in the case of oblique propagation of the wave with respect to the crystal axis, and only in the presence of free charge carriers. For an insulating uniaxial crystal the extraordinary wave is described by the dispersion relation

$$\omega = ck\left(\frac{\sin^2\theta}{\varepsilon_{\parallel}^{\mathrm{L}}} + \frac{\cos^2\theta}{\varepsilon_{\perp}^{\mathrm{L}}}\right)^{1/2} \tag{1.49}$$

This wave has neither cut-off nor resonance frequencies and there is no non-transparent region.

1.4 Phase and Group Velocities of the Waves

In this section we shall consider the peculiarities of the phase and group velocities of the electromagnetic waves in a uniaxial semiconductor. Those are determined by the well-known formulae [3]

$$\boldsymbol{v}_{\mathrm{p}} = \frac{\omega}{k}\boldsymbol{s} \tag{1.50}$$

$$\boldsymbol{v}_{\mathrm{g}} = \mathrm{d}\omega/\mathrm{d}\boldsymbol{k} \tag{1.51}$$

Using the dispersion relations obtained in the previous section, one finds that for the ordinary wave the directions of the vectors $\boldsymbol{v}_{\mathrm{p}}$ and $\boldsymbol{v}_{\mathrm{g}}$ coincide and the values are, respectively,

$$v_{\mathrm{p}} = c/\sqrt{\varepsilon_{\perp}(\omega)} \tag{1.52}$$

$$v_{\mathrm{g}} = c\sqrt{\varepsilon_{\perp}(\omega)}/\varepsilon_{\perp}^{\mathrm{L}} \tag{1.53}$$

These values are only real, naturally, inside the transparent region (1.42). Note that at the cut-off frequency $\omega = \omega_{0\perp}$ the phase velocity $v_{\mathrm{p}} \to \infty$ while $v_{\mathrm{g}} = 0$.

Unlike the ordinary wave, for the extraordinary one the group and phase velocity vectors are not parallel: v_{g} lies in the same xz-plane but makes an angle ϕ_{g} with the crystal axis, where

$$\tan\phi_{\mathrm{g}} = \frac{\varepsilon_{\perp}(\omega)}{\varepsilon_{\parallel}(\omega)}\tan\theta \tag{1.54}$$

It follows that $\boldsymbol{v}_{\mathrm{g}}$ is only parallel to $\boldsymbol{v}_{\mathrm{p}}$ ($\phi_{\mathrm{g}} = \theta$) at a single frequency

$$\omega = \omega_0 \sqrt{\frac{\mu_\perp - \mu_{||}}{\varepsilon_\perp^{\mathrm{L}} - \varepsilon_{||}^{\mathrm{L}}}} \tag{1.55}$$

at which $\varepsilon_\perp(\omega) = \varepsilon_{||}(\omega)$. Using Eqs. (1.43) and (1.54), we obtain

$$\tan\phi_{\mathrm{g}} = -\cot\phi_E \tag{1.56}$$

that is, $\boldsymbol{v}_{\mathrm{g}}$ is always perpendicular to the vector $\boldsymbol{E}$.

For a given value of ω the angle $\psi = |\theta - \phi_{\mathrm{g}}|$ between the phase and group velocity vectors as well as their absolute values are essentially dependent on the direction of propagation:

$$\cos\psi = \frac{\varepsilon_\perp(\omega)\sin^2\theta + \varepsilon_{||}(\omega)\cos^2\theta}{\sqrt{\varepsilon_\perp^2(\omega)\sin^2\theta + \varepsilon_{||}^2(\omega)\cos^2\theta}} \tag{1.57}$$

$$v_{\mathrm{p}} = c\sqrt{\frac{\sin^2\theta}{\varepsilon_{||}(\omega)} + \frac{\cos^2\theta}{\varepsilon_\perp(\omega)}} \tag{1.58}$$

$$v_{\mathrm{g}} = \frac{v_{\mathrm{p}}}{\cos\psi}\left[1 + \frac{\omega_0^2}{\omega^2}\,\frac{\varepsilon_\perp(\omega)\varepsilon_{||}^{-1}(\omega)\mu_{||}\sin^2\theta + \varepsilon_{||}(\omega)\varepsilon_\perp^{-1}(\omega)\mu_\perp\cos^2\theta}{\varepsilon_\perp(\omega)\sin^2\theta + \varepsilon_{||}(\omega)\cos^2\theta}\right]^{-1} \tag{1.59}$$

The regions of reality for v_{p} and v_{g} coincide with those given in Eqs. (1.45) and (1.46), which determine the transparent regions for the extraordinary wave.

Note that at $\omega \to \omega_{\mathrm{r}}$ where the resonance frequency ω_{r} is given by Eq. (1.47), both v_{p} and v_{g} vanish and the angle $\psi \to \pi/2$. At cut-off frequencies we obtain $v_{\mathrm{g}} \to 0$ and $v_{\mathrm{p}} \to \infty$; $\psi \to \theta$ for $\omega \to \omega_{0\perp}$ and $\psi \to \theta - \pi/2$ for $\omega \to \omega_{0||}$.

1.5 Longitudinal Plasmon Vibrations and Retardation Effect in Nonpolar Semiconductors

In this section we shall consider the spectrum of electromagnetic waves in a uniaxial semiconductor more profoundly and in detail. The waves discussed in previous sections can be considered as a result of taking into account the retardation effects in electronic plasma of the semiconductor. We shall see, in particular, that the splitting of the transparent region into two for the extraordinary wave is caused by coupling of the longitudinal electrostatic plasma vibrations with the electromagnetic waves propagating in nonconducting media.

Plasma vibrations in a uniaxial semiconductor are described by quasielectrostatic equations

$$\text{rot}\,\boldsymbol{E} = 0 \tag{1.60}$$

$$\text{div}\,\boldsymbol{D} = 0 \tag{1.61}$$

where $\boldsymbol{D}$ is the electric induction defined by Eq. (1.17). For plane waves Eq. (1.60) gives

$$\boldsymbol{E} = E\boldsymbol{k}/k \tag{1.60a}$$

i.e. the potential wave is longitudinal. The corresponding dispersion relation is easily found from Eqs. (1.61) and (1.18):

$$\varepsilon_{\alpha\beta}k_\alpha k_\beta = 0 \tag{1.61a}$$

For $\boldsymbol{k} = \{k_x, 0, k_z\}$ and $\varepsilon_{\alpha\beta}$ defined by Eq. (1.29), this relation gives

$$\varepsilon_\perp(\omega)\sin^2\theta + \varepsilon_{||}(\omega)\cos^2\theta = 0 \tag{1.62}$$

where θ is, as before, the angle between the wave vector and the crystal axis. From Eq. (1.62) we find the frequency of the plasma longitudinal vibrations, which is identical with the resonance frequency ω_r of the extraordinary wave defined by Eq. (1.47).

When the retardation of the Coulomb interaction of the charge carriers is taken into account, one must use Maxwell's equations (1.2) and (1.3) instead of Eqs. (1.60) and (1.61). Then, one obtains ordinary and extraordinary electromagnetic waves which are described by the dispersion relations (1.39) and (1.40), respectively.

There is only one branch of transversal electromagnetic vibrations whose frequency is given by Eq. (1.41). The corresponding dispersion curve is shown schematically in Fig. 1.1. For large values of the wave number $k >> \omega_0\sqrt{\mu_\perp}/c$ the function (1.41) behaves asymptotically as the frequency of transversal waves in nonconducting materials, which is given by Eq. (1.41a). For $k \to 0$ we obtain $\omega = \omega_{0\perp}$, which coincides with the frequency of plasma vibrations at $\theta = \pi/2$.
Using Eq. (1.40), we obtain two branches of TM-type vibrations with frequencies

$$\omega{\pm}^2 = \frac{1}{2}\left\{\omega_{0\perp}^2 + \omega_{0||}^2 + R_\theta c^2k^2 \pm \left[\left(\omega_{0\perp}^2 + \omega_{0||}^2 + R_\theta c^2k^2\right)^2 - 4\omega_\text{r}^2c^2k^2R_\theta\right]^{1/2}\right\} \tag{1.63}$$

where

$$R_\theta = \frac{\sin^2\theta}{\varepsilon_{||}^\text{L}} + \frac{\cos^2\theta}{\varepsilon_\perp^\text{L}}. \tag{1.64}$$

The corresponding dispersion curves are given in Fig. 1.2. We see that the wave can propagate in two frequency regions, one of which lies higher and the second lower than the plasma frequency ω_r. When $k \to \infty$, the high-frequency branch

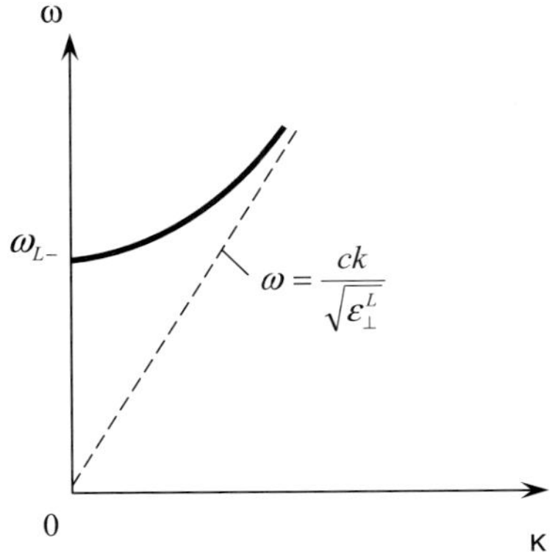

Fig. 1.1 Dispersion curve of TE-type mode.

approaches asymptotically the straight line “a” corresponding to the extraordinary wave in a nonconducting uniaxial crystal with frequency given in Eq. (1.49). The low-frequency branch approaches the frequency of longitudinal vibrations of plasma in a uniaxial semiconductor ω_{r}. In the long-wavelength limit $k \to \infty$ the frequencies of the branches in Fig. 1.2 coincide with the cut-off frequencies $\omega_{0\|}$ and $\omega_{0\perp}$. At intermediate values of the wave number k, the TM-polarized vibrations can be considered as the mixture of pure electromagnetic and plasma vibrations described by Eqs. (1.49) and (1.47), respectively. The coupling of the vibrations is especially strong in the vicinity of the crossing point of the corresponding dispersion curves represented in Fig. 1.2 by dotted lines.

Note that in an isotropic semiconductor both considered branches of TM waves are absent. The appearance of two frequency regions of transparency separated by

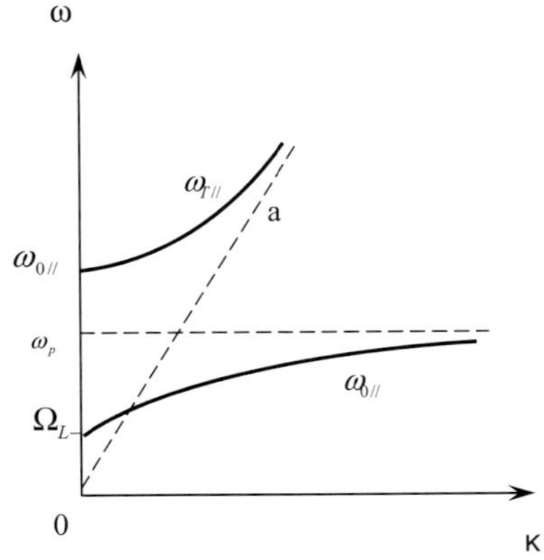

Fig. 1.2 Dispersion curves for TM-type modes. Straight line “a” represents the extraordinary wave described by Eq. (1.49).

the nontransparent region, as well as the possibility of the propagation of electromagnetic waves with frequencies lower than the plasma frequency, are two highly important effects of anisotropy. Both these effects take place only in the case of oblique propagation of the waves with respect to the crystal axis.

1.6
Long-Wavelength Optical Vibrations in Uniaxial Polar Crystals

In a polar crystal, the long-wavelength optical vibrations of the crystal lattice ($\lambda \gg a$, a-lattice constant) arise when ions with opposite signs of charge are displaced from their equilibrium positions. The vibrations are accompanied by high-frequency ($\omega \cong 3 \times 10^{13} s^{-1}$, infrared region) longitudinal electric fields [23]. The presence of such vibrations leads to the appearance of the frequency dependence of the crystal lattice dielectric permittivity tensor $\boldsymbol{\varepsilon}^{\mathrm{L}}$ and influences essentially the propagation of the infrared electromagnetic waves in the crystal.

To determine the kind of the functions $\varepsilon^{\mathrm{L}}_{\alpha\beta}(\omega)$ let us consider the simplest uniaxial polar crystal with two sublattices of ions with opposite signs of charge, for example CdS. The relative displacement of the sublattices, multiplied by the square root of the reduced mass per unit volume, will be represented by the vector $\boldsymbol{u}(\boldsymbol{r}, t)$. Then, the Lagrangian density for a lossless medium can be written in the form

$$L = \frac{1}{2}\left[(du/dt)^2 - a_{ij}u_iu_j + \varepsilon_0\chi_{ij}E_iE_j\right] - \beta_{ij}u_iE_j \tag{1.65}$$

where χ_{ij} is the electric susceptibility tensor describing polarization of the ions, and the components of the tensors a and β are coefficients describing mechanical quasielastic and electric forces, respectively. In a coordinate system with the z-axis parallel to the crystal axis, all these tensors are diagonal and have two different eigenvalues:

$$a_{xx} = a_{yy} = a_{\perp} \qquad a_{zz} = a_{\|} \qquad 1.66)$$

and the analogous expressions for the tensors β and χ.

Using the well-known general relation [32]

$$\frac{\mathrm{d}}{\mathrm{d}t}\left[\frac{\partial L}{\partial(\mathrm{d}u_i/\mathrm{d}t)}\right] - \frac{\partial L}{\partial u_i} = 0 \tag{1.67}$$

from Eq. (1.65) we obtain the equation of motion for the optical phonons in a uniaxial polar crystal:

$$\mathrm{d}^2\boldsymbol{u}/\mathrm{d}t^2 = -\boldsymbol{a}\boldsymbol{u} - \boldsymbol{\beta}\boldsymbol{E} \tag{1.68}$$

It is evident that the first term on the right-hand side of Eq. (1.68) is the local elastic restoring force, and the second is an additional restoring force arising due to the electric field associated with an optical lattice vibration.

In addition, the Lagrangian density (1.65) allows us to find the polarization density vector $\boldsymbol{P} = \partial L/\partial \boldsymbol{E}$. The result is

$$\boldsymbol{P} = -\boldsymbol{\beta u} + \varepsilon_0 \boldsymbol{\chi E} \tag{1.69}$$

The first term on the right-hand side of this equation is the polarization caused by the displacement of ions. The second term is caused by the polarization of the ions. Using Eq. (1.69), the electric flux density in the crystal lattice

$$\boldsymbol{D}_{\mathrm{L}} = \varepsilon_0 \boldsymbol{E} + \boldsymbol{P} \tag{1.70}$$

can be written in the form

$$\boldsymbol{D}_{\mathrm{L}} = \varepsilon_0 \varepsilon^{\infty} \boldsymbol{E} - \boldsymbol{\beta u} \tag{1.70a}$$

where

$$\varepsilon_{\alpha\beta}^{\infty} = \delta_{\alpha\beta} + \chi_{\alpha\beta}(\omega \to \infty) \tag{1.71}$$

is the crystal lattice dielectric permittivity tensor for such high (optical) frequencies, for which the vector $\boldsymbol{u} = 0$. It is obvious that this tensor coincides with the tensor $\boldsymbol{\varepsilon}_{\mathrm{L}}$ for a nonpolar crystal [see Eq. (1.4)].

Let us introduce now some new definitions:

$$\omega_{\mathrm{T}\perp} = \sqrt{a_{\perp}}, \quad \omega_{\mathrm{T}||} = \sqrt{a_{||}}, \tag{1.72}$$

$$\varepsilon_{\perp}^{0} = \varepsilon^{\infty}{}_{\perp} + \frac{\beta_{\perp}^{2}}{\varepsilon_0 a_{\perp}}, \quad \varepsilon_{||}^{0} = \varepsilon_{||}^{\infty} + \frac{\beta_{||}^{2}}{\varepsilon_0 a_{||}}, \tag{1.73}$$

$$\omega_{\mathrm{L}\perp} = \omega_{\mathrm{T}\perp}\sqrt{\varepsilon_{\perp}^{0}/\varepsilon_{\perp}^{\infty}}, \quad \omega_{\mathrm{L}||} = \sqrt{\varepsilon_{||}^{0}/\varepsilon_{||}^{\infty}}\,. \tag{1.74}$$

Assuming that $\boldsymbol{E}$, $\boldsymbol{u} \sim \exp[\mathrm{i}(k_x x + k_z z - \omega t)]$, using Eqs. (1.71)–(1.74) and eliminating the vector $\boldsymbol{u}$ from Eqs. (1.68) and (1.70a), we obtain

$$\boldsymbol{D}_{\mathrm{L}} = \varepsilon_0 \varepsilon_{\mathrm{L}}(\omega) \boldsymbol{E} \tag{1.75}$$

where

$$\varepsilon_{\mathrm{L}}(\omega) = \begin{pmatrix} \varepsilon_{\mathrm{L}\perp}(\omega) & 0 & 0 \\ 0 & \varepsilon_{\mathrm{L}\perp}(\omega) & 0 \\ 0 & 0 & \varepsilon_{||}(\omega) \end{pmatrix} \tag{1.76}$$

$$\varepsilon_{L\perp}(\omega) = \varepsilon_{\perp}^{\infty}\frac{\omega^2-\omega_{L\perp}^2}{\omega^2-\omega_{T\perp}^2}, \quad \varepsilon_{L||}(\omega) = \varepsilon_{||}^{\infty}\frac{\omega^2-\omega_{L||}^2}{\omega^2-\omega_{T||}^2}. \tag{1.77}$$

Using Eqs. (1.74) and (1.77), one finds that $\varepsilon_{||}^0$ and $\varepsilon_{\perp}^0$ are the static ($\omega = 0$) dielectric constants along and across the crystal axis. Before finding the meaning of the remaining notation, let us solve Eq. (1.68) in common with the quasielectrostatic equations rot $\boldsymbol{E} = 0$ and div $D_L = 0$. The results are two different modes of the crystal lattice optical vibrations:

- a transverse mode with frequency $\omega = \omega_{T\perp}$, for which

$$\boldsymbol{E} = 0 \text{ and } \boldsymbol{u} = \{0, \boldsymbol{u}, 0\} \tag{1.78}$$

- a longitudinal–transverse mode, for which the vector $\boldsymbol{u}$ lies in the xz-plane, so that both the longitudinal and the transverse (with respect to the wave vector $\boldsymbol{k}$) components exist. The vector $\boldsymbol{u}$ makes an angle φ_u with the crystal axis, where

$$\tan\varphi_u = \frac{\beta_{\perp}(\omega^2-\omega_{T||}^2)}{\beta_{||}(\omega^2-\omega_{T\perp}^2)}\tan^2\theta \tag{1.79}$$

The vibrations are accompanied by the longitudinal electric field

$$\boldsymbol{E} = \frac{\beta_{\perp}u_x\sin\theta + \beta_{||}u_z\cos\theta}{\varepsilon_{||}^{\infty}\,\varepsilon_{\perp}^{\infty}\,A(\theta)}\boldsymbol{k}/k \tag{1.80}$$

$$\text{where } A(\theta) = \frac{\sin^2\theta}{\varepsilon_{||}^{\infty}} + \frac{\cos^2\theta}{\varepsilon_{\perp}^{\infty}} \tag{1.81}$$

The frequencies of the optical vibrations are determined by the dispersion equation

$$\varepsilon_{L\perp}(\omega)\sin^2\theta + \varepsilon_{L||}(\omega)\cos^2\theta = 0 \tag{1.82}$$

which is similar to Eq. (1.62), but, unlike that, it has two different solutions for ω^2:

$$\omega_{\pm}^2(\theta) = \left[B(\theta) \pm \sqrt{B^2(\theta) - 4(\omega_{L\perp}\omega_{L||})^2A_0(\theta)A(\theta)}\right]/2A(\theta) \tag{1.83}$$

where

$$B(\theta) = B_1(\theta)\omega_{L||}^2 + B_2(\theta)\omega_{L\perp}^2 \tag{1.83a}$$

$$A_0(\theta) = \frac{\sin^2\theta}{\varepsilon_{||}^0} + \frac{\cos^2(\theta)}{\varepsilon_{\perp}^0} \tag{1.84}$$

$$B_1(\theta) = \frac{\sin^2(\theta)}{\varepsilon_{||}^0} + \frac{\cos^2(\theta)}{\varepsilon_{\perp}^{\infty}} \qquad B_2(\theta) = \frac{\sin^2(\theta)}{\varepsilon_{||}^{\infty}} + \frac{\cos^2(\theta)}{\varepsilon_{\perp}^0} \tag{1.85}$$

An example of the angle dependence of the functions $\omega \pm(\theta)$ is shown schematically in Fig. 1.3 for a crystal with characteristic frequencies $\omega_{\mathrm{T}\perp} < \omega_{\mathrm{T}||} < \omega_{\mathrm{L}\perp} < \omega_{\mathrm{L}||}$. Note that for $\theta = 0$ and $\theta = \pi/2$, one of the solutions of Eq. (1.82) corresponds to the longitudinal ($\boldsymbol{u}||\boldsymbol{k}$) mode and the second to the transverse mode ($\boldsymbol{u}\perp\boldsymbol{k}$). The frequencies of the transverse modes are equal to $\omega_{\mathrm{T}\perp}$ for $\theta = 0$ and $\omega_{\mathrm{T}||}$ for $\theta = \pi/2$. Those for the longitudinal modes are $\omega_{\mathrm{L}||}$ for $\theta = 0$ and $\omega_{\mathrm{L}\perp}$ for $\theta = \pi/2$. Hence, the first index in the notation $\omega_{\mathrm{T}\perp}, \omega_{\mathrm{T}||}, \omega_{\mathrm{L}\perp}$ and $\omega_{\mathrm{L}||}$ characterizes the orientation of the displacement vector $\boldsymbol{u}$ with respect to the wave vector $\boldsymbol{k}$ while the second index is with respect to the crystal axis. In the case of isotropic (cubic) crystals $\omega_{\mathrm{T}\perp} = \omega_{\mathrm{T}||}$, $\omega_{\mathrm{L}\perp} = \omega_{\mathrm{L}||}$ and the longitudinal and transverse vibrations are split for an arbitrary direction of propagation.

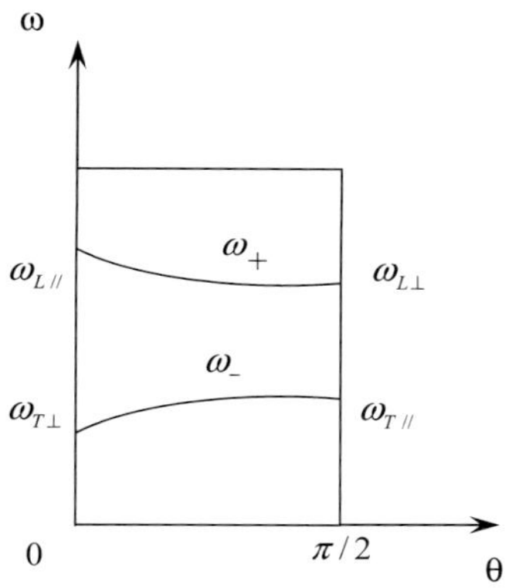

Fig. 1.3 Angle dependence of the frequencies for longitudinal–transverse modes described by Eq. (1.83).

Thus, the presence of the optical axis in a crystal leads to the doubling of characteristic frequencies of the optical phonons. In addition, it leads to the appearance of the mixed longitudinal–transverse modes whose frequencies are given in Eq. (1.83) and depend essentially on the direction of propagation.

2
Bulk Polaritons in Uniaxial Polar Semiconductors

2.1
Retardation Effects in Nonconducting Polar Crystals. Dispersion Relations for Phonon–Polaritons

In this section we shall consider the behavior of electromagnetic waves in a polar uniaxial crystal taking into account the interaction between the waves and the optical lattice vibrations. In other words, our task is to find the retardation effect on the long-wavelength optical phonons. It is well known that in the isotropic case this interaction leads to the appearance of the coupled waves, the so-called phonon–polaritons [33].

To find the dispersion relations describing phonon–polaritons in a uniaxial crystal, let us first eliminate the relative displacement vector $\boldsymbol{u}$ from the equation of motion (1.68) and the constitutive relation (1.70a). Then, the investigation of the polaritons reduces to the solution of Maxwell's equations (1.2) and (1.3) with the vector $\boldsymbol{D}_{\mathrm{L}}$ given by Eq. (1.75). In a Cartesian system of coordinates with the z-axis parallel to the crystal optical axis, for the plane waves $\boldsymbol{E}$, $\boldsymbol{H}$, $\boldsymbol{u}$ ~$\exp[\mathrm{i}(k_x x + k_z z - \omega t)]$ we obtain the following system of the wave equations:

$$\left[k_x^2 + k_z^2 - \frac{\omega^2}{c^2}\varepsilon_\perp^{\mathrm{L}}(\omega)\right]E_y = 0,$$

$$\left[k_z^2 - \frac{\omega^2}{c^2}\varepsilon_{\mathrm{L}\perp}(\omega)\right]E_x - k_x k_z E_z = 0, \tag{2.1}$$

$$k_x k_z E_x - \left[k_x^2 - \frac{\omega^2}{c^2}\varepsilon_{\mathrm{L}\|}(\omega)\right]E_z = 0,$$

where $\varepsilon_{\mathrm{L}\perp,\|}(\omega)$ are given by Eq. (1.77). The dispersion equation corresponding to Eqs. (2.1) has two different solutions for the refractive index $n = ck/\omega$:

$$n_{\mathrm{o}}^2 = \varepsilon_{\mathrm{L}\perp}(\omega) \tag{2.2}$$

and

$$n_{\mathrm{e}}^2 = \frac{\varepsilon_{\mathrm{L}\perp}(\omega)\varepsilon_{\mathrm{L}\|}(\omega)}{\varepsilon_{\mathrm{L}\perp}(\omega)\sin^2\theta + \varepsilon_{\mathrm{L}\|}(\omega)\cos^2\theta}. \tag{2.3}$$

Radiowaves and Polaritons in Anisotropic Media. Roland H. Tarkhanyan and Nikolaos K. Uzunoglu

ISBN: 3-527-40615-8

The first solution corresponds to the ordinary transverse wave, for which $\boldsymbol{E}\|\boldsymbol{u}\|oy$. Usually this wave is called a TE-polarized one. The second solution corresponds to the extraordinary longitudinal–transverse waves (for $\theta \neq 0, \pi/2$) in which the vectors $\boldsymbol{E}$, $\boldsymbol{u}$ (and $\boldsymbol{k}$) lie in the xz-plane, perpendicular to the magnetic field vector $\boldsymbol{H}$. That is why this wave is usually called a TM-polarized one.

Let us consider first the transverse wave. It can be propagated ($n_o^2 > 0$) in two frequency regions: $\omega < \omega_{T\perp}$ and $\omega > \omega_{L\perp}$. The frequency $\omega = \omega_{L\perp}$ corresponds to the cut-off ($n_o = 0$), the limit $\omega = \omega_{T\perp}$ corresponds to the resonance ($n_o \rightarrow \infty$) and the remaining range of frequencies $\omega_{T\perp} < \omega < \omega_{L\perp}$ corresponds to the absorption ($n_o^2 < 0$).

For a given $\boldsymbol{k}$, there are two transverse branches whose frequencies are given by

$$\omega_{T\pm} = \frac{1}{\sqrt{2}} \left[\Omega^2 \pm \left(\Omega^4 - 4\omega_{T\perp}^2 c^2 k^2 / \varepsilon_\perp^\infty \right)^{1/2} \right]^{1/2} \tag{2.4}$$

where

$$\Omega^2 \equiv \omega_{L\perp}^2 + c^2 k^2 / \varepsilon_\perp^\infty \tag{2.5}$$

In the limit $k \rightarrow 0$, we obtain $\omega_{T+} \approx \omega_{L\perp}$ and $\omega_{T-} \approx ck/\sqrt{\varepsilon_\perp^0}$, that is, the low-frequency branch approaches the transverse electromagnetic wave given by Eq. (1.41a), where $\varepsilon_\perp^L$ should be replaced by $\varepsilon_\perp^0$ given in Eq. (1.73). As to the high-frequency branch, it corresponds to the longitudinal optical phonons at $\theta = \pi/2$. For large values of the wave number k, we obtain $\omega_{T+} \approx ck/\sqrt{\varepsilon_\perp^\infty}$ and $\omega_{T-} \approx \omega_{T\perp}$, where the latter is the frequency of the transverse optical phonons.

For $\boldsymbol{k}\|oz$ ($\theta = 0$), the second solution (2.3) splits in two: one of them corresponds to the transverse wave with $\boldsymbol{E}\|ox$ and $n_e = n_o$, and the second to the longitudinal wave of frequency $\omega_{L\|}$, for which $\boldsymbol{E}\|\boldsymbol{u}\|oz$. In this case, as in the isotropic case, the electromagnetic wave is transversal and is only coupled to the transverse optical phonons.

In the limit $\theta = \pi/2$ the solution in question splits into a longitudinal wave ($\boldsymbol{E}\|\boldsymbol{u}\|ox$) of frequency $\omega_{L\perp}$ and a transverse wave ($\boldsymbol{E}\|\boldsymbol{u}\|oz$) with the refractive index $n^2 = \varepsilon_\|^L(\omega)$. The latter solution can be obtained from Eq. (2.2) if the symbol $\perp$ is replaced by the symbol $\|$ in all the relevant equations.

We have to note that for $\theta = 0$ the absorption regions of two transverse waves with different polarizations are identical ($\omega_{T\perp} < \omega < \omega_{L\perp}$), while for $\theta = \pi/2$ they are different.

In particular, for $\omega_{T\|} > \omega_{L\perp}$ and $\omega_{T\perp} > \omega_{L\|}$ the transparency regions of the transverse waves with $n^2 = \varepsilon_{L\|}(\omega)$ and $n^2 = \varepsilon_{L\perp}(\omega)$ overlap, which implies that the crystal is transparent in the whole frequency range.

2.2 Dispersion of Longitudinal–Transverse Phonon–Polaritons

For $0 < \theta < \pi/2$ the relation given by Eq. (2.3) describes the mixed longitudinal–-transverse waves, which cannot be propagated in an isotropic medium. For a given value of the wave vector $\boldsymbol{k}$, there are three branches corresponding to a "mixture" of two phonon branches, whose frequencies $\omega_{\pm}(\theta)$ are given by Eq. (1.83), with the extraordinary electromagnetic wave described by the dispersion relation

$$\omega = ck\sqrt{A(\theta)} \tag{2.6}$$

where $A(\theta)$ is defined by Eq. (1.81). [See also Eq. (1.49).] Substituting Eq. (1.77) into Eq. (2.3), we obtain

$$n_e^2 = \frac{(\omega^2 - \omega_{L\perp}^2)(\omega^2 - \omega_{L\|}^2)}{[\omega^2 - \omega_+^2(\theta)][\omega^2 - \omega_-^2(\theta)]A(\theta)}. \tag{2.7}$$

The wave is cut off at $\omega = \omega_{L\perp}$ and $\omega = \omega_{L\|}$; it has two resonance frequencies at $\omega_+(\theta)$ and $\omega_-(\theta)$, and two absorption regions, where $n_e^2 < 0$. For $\omega_{L\perp} < \omega_{L\|}$, these regions are given by

$$\omega_-(\theta) < \omega < \omega_{L\perp}, \quad \omega_+(\theta) < \omega < \omega_{L\|}. \tag{2.8}$$

The frequencies of the coupled phonon–polariton branches correspond to the solutions of the following bicubic equation:

$$\omega^2(\omega^2 - \omega_{L\perp}^2)(\omega^2 - \omega_{L\|}^2) = c^2k^2[A(\theta)\omega^4 - \omega^2(B_1\omega_{L\|}^2 + B_2\omega_{L\perp}^2) + A_0\omega_{L\|}^2\omega_{L\perp}^2], \tag{2.9}$$

where $A_0 = A_0(\theta)$ and $B_{1,2} = B_{1,2}(\theta)$ are defined in Eqs. (1.84) and (1.85), respectively.

In Fig. 2.1 the dispersion curves $\omega(k)$ are shown schematically, for a given value of θ. In the limit $k \to 0$ the frequencies of the coupled waves are given by

$$\omega_1 \cong ck\sqrt{A_0(\theta)}, \tag{2.10}$$

$$\omega_2 \cong \omega_{L\perp}\left[1 + \left(\frac{1}{\varepsilon_\perp^\infty} - \frac{1}{\varepsilon_\perp^0}\right)\frac{c^2k^2}{\omega_{L\perp}^2}\cos^2\theta\right], \tag{2.11}$$

$$\omega_3 \cong \omega_{L\|}\left[1 + \left(\frac{1}{\varepsilon_\|^\infty} - \frac{1}{\varepsilon_\|^0}\right)\frac{c^2k^2}{\omega_{L\|}^2}\sin^2\theta\right]. \tag{2.12}$$

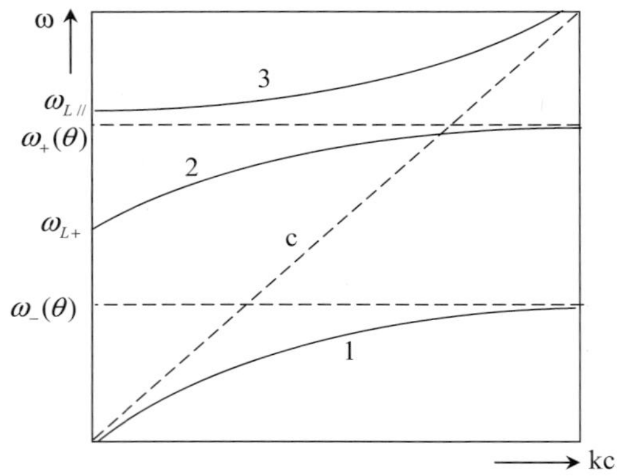

Fig. 2.1 Dispersion curves for the coupled longitudinal–transverse phonon–polaritons. Line "c" corresponds to the extraordinary electromagnetic wave described by Eq. (2.6).

Therefore, the long-wavelength limit of the lower branch corresponds to the extraordinary electromagnetic wave [$A(\theta)$ in Eq. (2.6) should be replaced by$A_0(\theta)$], and the two upper branches correspond to the longitudinal optical phonons which travel in the directions parallel and perpendicular to the crystal axis.

In the limit $k \to \infty$, the solutions of Eq. (2.9) have the form

$$\omega_1 \cong \omega_-(\theta), \tag{2.13}$$

$$\omega_2 \cong \omega_+(\theta), \tag{2.14}$$

$$\omega_3 \cong ck\sqrt{A(\theta)}. \tag{2.15}$$

The upper branch approaches asymptotically the longitudinal–transverse branch of the electromagnetic waves defined by Eq. (2.6), and the two lower branches approach the longitudinal–transverse phonon modes.

2.3 Dielectric Permittivity Tensor for Uniaxial Polar Semiconductors. Coupling of Plasmons and Optical Phonons

In this section we shall investigate the normal oscillations of two interacting oscillating systems in a uniaxial polar semiconductor in the absence of retardation effects: plasma oscillations and long-wavelength optical phonons. If the frequency of the electronic plasma longitudinal oscillations is comparable to the characteristic frequencies of the longitudinal optical phonons, the coupled phonon–plasmon vibrations appear which for isotropic (cubic) semiconductors have been considered in Ref. [26]. These vibrations were observed in the Raman light scattering spectrum of GaAs [27, 35] and in the infrared reflection spectra of GaP [36], GaAs [37], InSb [38] and GaSb [39]. In all these crystals two branches of the coupled longitudinal vibrations were observed for which $\boldsymbol{u} \| \boldsymbol{E} \| \boldsymbol{k}$.

To study the peculiarities of the mixed phonon–plasmon vibrations in a uniaxial polar semiconductor, we will use the quasistatic equations (1.60) and (1.61), Eqs. (1.68) and (1.69) for the optical vibrations and polarization density vector, and the linearized kinetic equation (1.9) for the conduction electrons with ellipsoidal iso-energetic surfaces. The simultaneous solution of these equations yields

$$E_i = \frac{k_i k_l \beta_{jl} u_j}{\varepsilon_{\alpha\beta} k_\alpha k_\beta}, \tag{2.16}$$

$$\omega^2 u_i = \left(a_{ik} + \frac{\beta_{il} \beta_{kj} k_l k_j}{\varepsilon_{\alpha\beta} k_\alpha k_\beta} \right) u_k, \tag{2.17}$$

where $\varepsilon_{\alpha\beta}$ is defined by Eq. (1.19). For $\boldsymbol{k} = \{k_x, 0, k_z\}$, Eq. (2.17) yields one transverse phonon branch with the former frequency $\omega = \omega_{\mathrm{T}\perp}$, $\boldsymbol{E} = 0$ and $\boldsymbol{u} \| oy$, and three new branches corresponding to the coupled phonon–plasmon vibrations, for which vector $\boldsymbol{u}$ lies in the xz-plane and has both longitudinal and transverse components with respect to the wave vector, and which are accompanied by the longitudinal electric field defined in Eq. (2.15). The best way to find the frequencies of such waves is the following. First of all, for the total electric flux density we find

$$\boldsymbol{D} = \varepsilon_0 \left(\varepsilon^\infty + \frac{\mathrm{i}}{\varepsilon_0 \omega} \boldsymbol{\sigma} \right) \boldsymbol{E} - \boldsymbol{\beta} \boldsymbol{u}. \tag{2.18}$$

Eliminating $\boldsymbol{u}$ from Eqs. (2.17) and (2.18) and substituting $\boldsymbol{\sigma}$ given by Eq. (1.16a), for the nonzero components of the complex dielectric permittivity tensor we obtain

$$\varepsilon_{xx} = \varepsilon_{yy} \equiv \varepsilon_\perp = \varepsilon_\perp^\infty \left(\frac{\omega^2 - \omega_{\mathrm{L}\perp}^2}{\omega^2 - \omega_{\mathrm{T}\perp}^2} - \frac{\omega_{0\perp}^2}{\omega^2} \right), \varepsilon_{zz} \equiv \varepsilon_{\|} = \varepsilon_{\|}^\infty \left(\frac{\omega^2 - \omega_{\mathrm{L}\|}^2}{\omega^2 - \omega_{\mathrm{T}\|}^2} - \frac{\omega_{0\|}^2}{\omega^2} \right). \tag{2.19}$$

Then, the general dispersion relation (1.61) for the potential waves leads again to Eq. (1.62), which after substitution of Eqs. (2.19) gives the following bicubic equation for ω:

$$\frac{\omega_{\mathrm{r}}^2}{\omega^2} = \frac{\left[\omega^2 - \omega_+^2(\theta)\right]\left[\omega^2 - \omega_-^2(\theta)\right]}{\left(\omega^2 - \omega_{\mathrm{T}\perp}^2\right)\left(\omega^2 - \omega_{\mathrm{T}\|}^2\right)}. \tag{2.20}$$

Here ω_{r} and $\omega_\pm(\theta)$ are given by Eqs. (1.47) and (1.83), respectively. It can easily be seen that all three solutions of Eq. (2.20), which correspond to three branches of the coupled phonon–plasmon vibrations, are real. In Fig. 2.2 the dependence of these branches on the free charge carrier density $N \approx \omega_r^2$ is shown, for a given value of $\theta \neq 0,\ \pi/2$. At high density, when the plasma frequency is higher than all the characteristic lattice frequencies, the upper branch approaches asymptotically the straight line "a", which corresponds to the plasma oscillations defined in Eq. (1.47). The remaining branches 1 and 2 approach the frequencies of the transverse optical phonons. For $\omega_r << \omega_{\mathrm{T}\perp,\|}$, the lower branch is essentially plasma-like, and the two upper branches correspond to the longitudinal–transverse optical vibrations with a small component of plasma oscillations. The coupling of plasmons and phonons is especially strong near the crossing points C_1 and C_2 of the dotted lines.

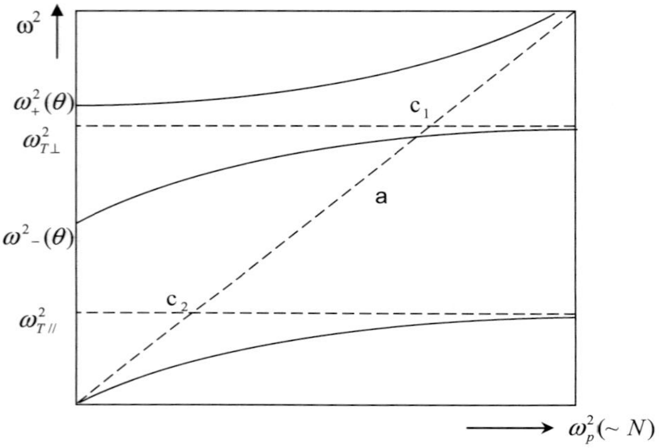

Fig. 2.2 The dependence of ω^2 on N for longitudinal–transverse phonon–plasmon branches.

As has been mentioned above, there are only two branches of the coupled phonon–plasmon vibrations in an isotropic material. For a uniaxial crystal an analogous situation occurs only at the wave propagation along and across the crystal axis. Then, one of the roots of Eq. (2.20) corresponds to the transverse optical phonons and the remaining two branches become purely longitudinal. In the limit $\theta = 0$, Eq. (2.20) yields $\omega = \omega_{\mathrm{T}}$ for the transverse branch with $\boldsymbol{u}\|0x$ and

$$\omega_{\mathrm{L}\pm}^2 = \frac{1}{2}\left(\omega_{\mathrm{L}\|}^2 + \omega_{0\|}^2\right) \pm \frac{1}{2}\left[\left(\omega_{\mathrm{L}\|}^2 + \omega_{0\|}^2\right)^2 - 4\omega_{\mathrm{T}\|}^2\omega_{0\|}^2\right]^{1/2} \tag{2.21}$$

for the coupled longitudinal branches in which $\boldsymbol{u}|oz$. In the limit $\theta = \pi/2$ we obtain $\omega = \omega_{\mathrm{T}\|}$ for the transverse branch ($\boldsymbol{u}\|oz$) and

$$\Omega_{\mathrm{L}\pm}^2 = \frac{1}{2}\left(\omega_{\mathrm{L}\perp}^2 + \omega_{0\perp}^2\right) \pm \frac{1}{2}\left[\left(\omega_{\mathrm{L}\perp}^2 + \omega_{0\perp}^2\right)^2 - 4\omega_{\mathrm{T}\perp}^2\omega_{0\perp}^2\right]^{1/2} \tag{2.22}$$

for the longitudinal branches with $\boldsymbol{u}\|ox$.

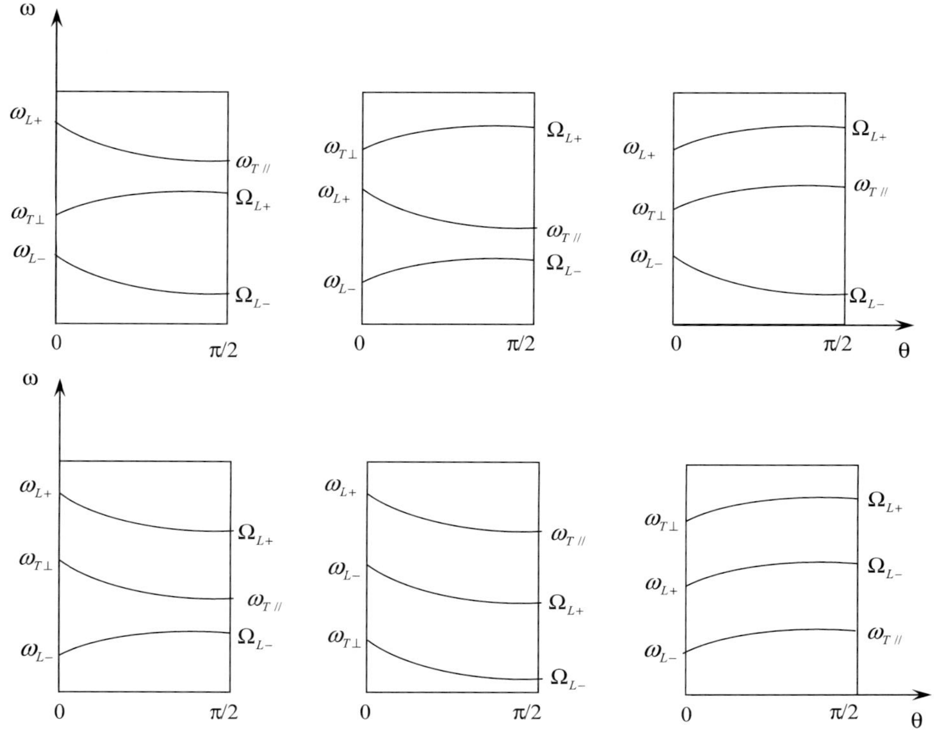

Fig. 2.3 The dependence of ω on θ for longitudinal–transverse phonon–plasmon branches (θ is the angle between the wave vector and the crystal axis).

For $\theta \neq 0,\ \pi/2$, the anisotropy leads not only to the increasing of the quantity of coupled branches in the spectrum of phonon–plasmon vibrations. It also causes the appearance of the essential dependence of the frequencies on the direction of propagation. Six different versions of such dependence are shown in Fig. 2.3. The frequencies of the coupled vibrations increase or decrease with increasing θ, in dependence on the relative magnitudes of the characteristic frequencies $\omega_{\mathrm{T}\perp}$, $\omega_{\mathrm{T}\|}$, $\omega_{\mathrm{L}\pm}$, and $\Omega_{\mathrm{L}\pm}$. It is evident that such a diverse behavior of the frequencies in dependence on the direction of propagation is conditioned by the presence of the anisotropy and is impossible in an isotropic medium.

2.4 Coupling of Electromagnetic and Phonon–Plasmon Vibrations

It is now time to consider the propagation of the electromagnetic waves that interact simultaneously with the plasma and lattice vibrations. We shall see that the mutual interaction of these three vibrational subsystems and the anisotropy can modify considerably the wave propagation compared with the situation in an isotropic medium or with systems where the electromagnetic waves interact with only one of the aforementioned subsystems.

A simultaneous solution of the Maxwell equations (1.2a) and (1.3), containing the electric flux density vector given by Eq. (2.18), and of the equations of motion for the optical phonons (1.68) and for the free charge carriers whose energetic spectrum is given by Eq. (1.1), in the collisionless situation yields the dispersion relation for the coupled waves (1.37), where the diagonal permittivity tensor is given by Eqs. (2.19). Using Eq. (1.37), we obtain an ordinary TE-polarized wave with refractive index

$$n_o^2 = \varepsilon_{yy} = \varepsilon_\perp^\infty \frac{(\omega^2 - \Omega_{L-}^2)(\omega^2 - \Omega_{L+}^2)}{\omega^2(\omega^2 - \omega_{T\perp}^2)}, \tag{2.23}$$

and an extraordinary TM-polarized wave whose refractive index depends on the direction of propagation:

$$n_e^2 = \frac{(\omega^2 - \omega_{L-}^2)(\omega^2 - \omega_{L+}^2)(\omega^2 - \Omega_{L-}^2)(\omega^2 - \Omega_{L+}^2)}{\omega^2\left[\left(\omega^2 - \omega_{T\|}^2\right)(\omega^2 - \Omega_{L-}^2)(\omega^2 - \Omega_{L+}^2)\frac{\sin^2\theta}{\varepsilon_\|^\infty} + \frac{\cos^2\theta}{\varepsilon_\perp^\infty}(\omega^2 - \omega_{T\perp}^2)(\omega^2 - \omega_{L-}^2)(\omega^2 - \omega_{L+}^2)\right]} \tag{2.24}$$

The ordinary wave exists ($n_o^2 > 0$) in the following two frequency regions:

$$\Omega_{L-} < \omega < \omega_{T\perp} \qquad \omega > \Omega_{L+} \tag{2.25}$$

Cut off occurs at $\omega = \Omega_{L-}$ and$\omega = \Omega_{L+}$, and a resonance at $\omega = \omega_{T\perp}$. In the latter case, the absorption should be taken into account, that is, the friction term $\gamma\, d\boldsymbol{u}/dt$ should be added to Eq. (1.68), and then we obtain the following expression for the refractive index at the resonance:

$$n_o^2 = \varepsilon_\perp^\infty \left(1 - \frac{\omega_r^2}{\omega_{T\perp}^2}\right) - i\frac{\omega_{T\perp}}{\gamma}\left(\varepsilon_\perp^0 - \varepsilon_\perp^\infty\right). \tag{2.26}$$

Therefore, an allowance for the lattice vibrations modifies considerably the properties of the transverse wave with dispersion relation (1.41); in particular, a resonance occurs at the frequency of transverse optical phonons. Moreover, the low-frequency boundary of the region in which the wave exists is modified considerably (Ω_{L+} replaces $\omega_{0\perp}$) and, finally, an additional transparency region is created.

The properties of the ordinary wave are also considerably different from those of the transverse phonon–polaritons studied in Section 2.1: an additional cut-off occurs, the boundary of the high-frequency transparency region is modified (Ω_{L+} replaces $\omega_{0\perp}$) and, finally, there are now two absorption regions instead of one in the case of phonon–polaritons. It follows that the question is about a new type of collective vibrations, which can be called "phonon–plasmon polaritons".

For a given wave vector $\boldsymbol{k}$ there are two branches of those coupled oscillations whose frequencies are given by

$$\omega^2 = \frac{1}{2}\left\{\omega_{L\perp}^2 + \omega_{0\perp}^2 + \frac{c^2k^2}{\varepsilon_\perp^\infty} \pm \left[\left(\omega_{L\perp}^2 + \omega_{0\perp}^2 + \frac{c^2k^2}{\varepsilon_\perp^\infty}\right)^2 - 4\omega_{T\perp}^2\left(\omega_{0\perp}^2 + \frac{c^2k^2}{\varepsilon_\perp^\infty}\right)\right]^{1/2}\right\} \tag{2.27}$$

and are shown schematically in Fig. 2.4 as functions of $ck(\varepsilon_\perp^\infty)^{-1/2}$. The absorption region $\omega < \Omega_{L-}$ is due to the free carriers, whereas the second absorption region $\omega_{T\perp} < \omega < \omega_{L\perp}$ arises as a result of screening of the electromagnetic waves by the lattice vibrations. In the limit $ck \to \infty$, the high-frequency branch approaches asymptotically the straight line $\omega = ck(\varepsilon_\perp^\infty)^{-1/2}$, which corresponds to the transverse electromagnetic waves in an insulating crystal (the dashed line); the frequency of the lower branch approaches the frequency of the transverse optical phonons $\omega_{T\perp}$. For $k \to 0$, the frequencies of the branches approach $\Omega_{L\pm}$ corresponding to the longitudinal phonon–plasmon branches at $\theta = \pi/2$.

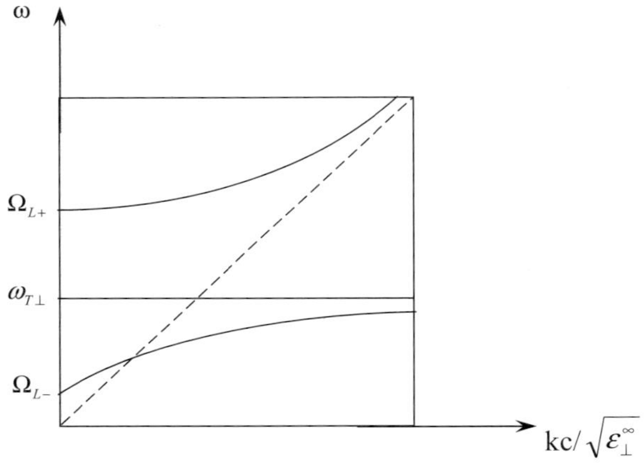

Fig. 2.4 Dispersion curves for transverse branches of phonon–plasmon polaritons given by Eq. (2.27).

The dependence of the frequencies defined by Eq. (2.27) on the conduction electron density $N \sim \omega_{0\perp}^2$ can be shown in Fig. 2.4, if $ck(\varepsilon_\perp^\infty)^{-1/2}$ on the abscissa is replaced by $\omega_{0\perp}$ and $\Omega_{L+,-}$ on the ordinate is replaced by $\omega_{T+,-}$ given by Eq. (2.4). In this case, the dashed line corresponds to the plasma oscillations traveling in the direction perpendicular to the crystal axis with frequency $\omega_{0\perp}$.

2.5
Spectrum of Extraordinary Phonon–Plasmon Polaritons

As we have seen in previous sections, the presence of an anisotropy leads to the creation of extraordinary waves that cannot propagate in an isotropic medium. In this section we will extend our discussion to extraordinary phonon–plasmon polaritons which are described by the dispersion equation (2.24). The wave exhibits its four cut-offs at $\omega = \Omega_{L\pm}$ and $\omega = \omega_{L\pm}$ corresponding to the longitudinal phonon–plasmon oscillations, and three resonances at the frequencies of longitudinal–transverse phonon–plasmon vibrations given by Eq. (2.20) and studied in detail in Section 2.3. Therefore, the interaction of the electromagnetic waves and the plasma oscillations with optical lattice vibrations gives rise to two additional resonances and two cut-offs; this effect creates two additional transparency regions and two absorption regions. Thus, there are four regions of either type. For $\theta = 0$ and $\theta = \pi/2$, the wave splits into one transverse and two longitudinal waves. The refractive index of the transverse wave is identical with Eq. (2.23) for $\theta = 0$ and is given by

$$n^2 = \varepsilon_{\|}^{\infty} \frac{(\omega^2 - \omega_{L+}^2)(\omega^2 - \omega_{L-}^2)}{\omega^2\left(\omega^2 - \omega_{T\|}^2\right)} \tag{2.28}$$

for $\theta = \pi/2$. The wave governed by Eq. (2.28) has all the properties of the wave described by Eq. (2.23) if all the quantities labeled by $\perp$ are replaced by the corresponding quantities with the symbol $\|$. It should be noted that the vector $\boldsymbol{E}$ corresponding to the wave described by Eq. (2.23) is parallel to the y-axis, that is, it is orthogonal to the plane of propagation, while it is parallel to the x-axis in the wave corresponding to $\theta = 0$. For the wave governed by Eq. (2.28), the vector $\boldsymbol{E}$ is parallel to the z-axis.

The frequencies of the longitudinal branches are equal to ω_{L+} and ω_{L-} for $\theta = 0$ and to Ω_{L+} and Ω_{L-} for $\theta = \pi/2$. Note that in the absence of the optical phonons, that is, for a nonpolar crystal, there is always only one longitudinal branch at $\theta = 0$ and $\theta = \pi/2$; its frequency is, respectively, $\omega_{0\|}$ and $\omega_{0\perp}$, which represent the limiting values of ω_{L+} and Ω_{L+} for very high plasma densities, when the plasma frequency is much higher than all the characteristic lattice frequencies.

For $\theta \neq 0$, $\pi/2$ and for a given value of $\boldsymbol{k}$, there are four branches of the extraordinary phonon–plasmon polaritons, which do not appear in the case of isotropic (cubic) crystals and represent a "mixture" of the electromagnetic waves with the optical phonons and plasmons. In Fig. 2.5, the dependences of these branches' frequencies on the wave number k are shown schematically, for the case when $\Omega_{L-} < \omega_{L-} < \Omega_{L+} < \omega_{L+}$. We see that for large values of k, the frequencies of the three low-frequency branches approach the resonance frequencies which correspond to the roots of Eq. (2.20) (dashed lines), and the high-frequency branch approaches a straight line corresponding to the longitudinal–transverse electromagnetic wave in a uniaxial insulating crystal with a rigid lattice described by Eq. (1.49). For $k \to 0$, the frequencies of the branches are given by

$$\omega_{1,2} \approx \Omega_{L\pm} \left[1 + \frac{c^2 k^2 \left(\Omega_{L\pm}^2 - \omega_{T\perp}^2 \right) \cos^2 \theta}{2\varepsilon_{\perp}^{\infty} \Omega_{L\pm}^2 \left(\Omega_{L+}^2 - \Omega_{L-}^2 \right)} \right], \tag{2.29}$$

$$\omega_{3,4} \approx \omega_{L\pm} \left[1 + \frac{c^2 k^2 \left(\omega_{L\pm}^2 - \omega_{T\|}^2 \right) \sin^2 \theta}{2\varepsilon_{\|}^{\infty} \omega_{L\pm}^2 \left(\omega_{L+}^2 - \omega_{L-}^2 \right)} \right], \tag{2.30}$$

that is, they are closed to the frequencies of the phonon–plasmon branches. There are four transparency regions; three of them (low-frequency regions) are bounded from above by the resonance frequencies which depend essentially on the angle θ, and from below by the cut-off frequency ω_{L+} or Ω_{L+}, depending on their relative magnitude. Twenty different mutual positions of the transparency and absorption regions are possible; the corresponding dependences on θ are shown in Fig. 2.6, for three alternatives.

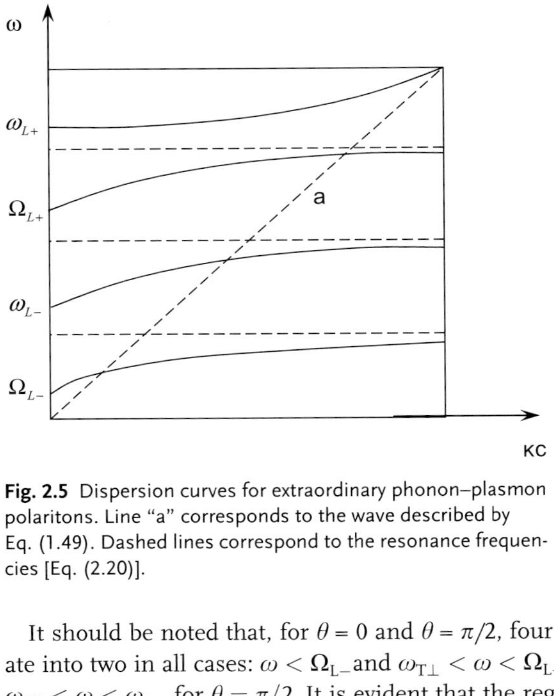

Fig. 2.5 Dispersion curves for extraordinary phonon–plasmon polaritons. Line "a" corresponds to the wave described by Eq. (1.49). Dashed lines correspond to the resonance frequencies [Eq. (2.20)].

It should be noted that, for $\theta = 0$ and $\theta = \pi/2$, four absorption regions degenerate into two in all cases: $\omega < \Omega_{L-}$ and $\omega_{T\perp} < \omega < \Omega_{L+}$ for $\theta = 0$ and $\omega < \omega_{L-}$ and $\omega_{T\|} < \omega < \omega_{L+}$ for $\theta = \pi/2$. It is evident that the regions of total reflection corresponding to the normal incidence should reduce to one of the alternatives shown in Fig. 2.6, depending on the angle θ between the crystal axis and the normal to the reflecting surface.

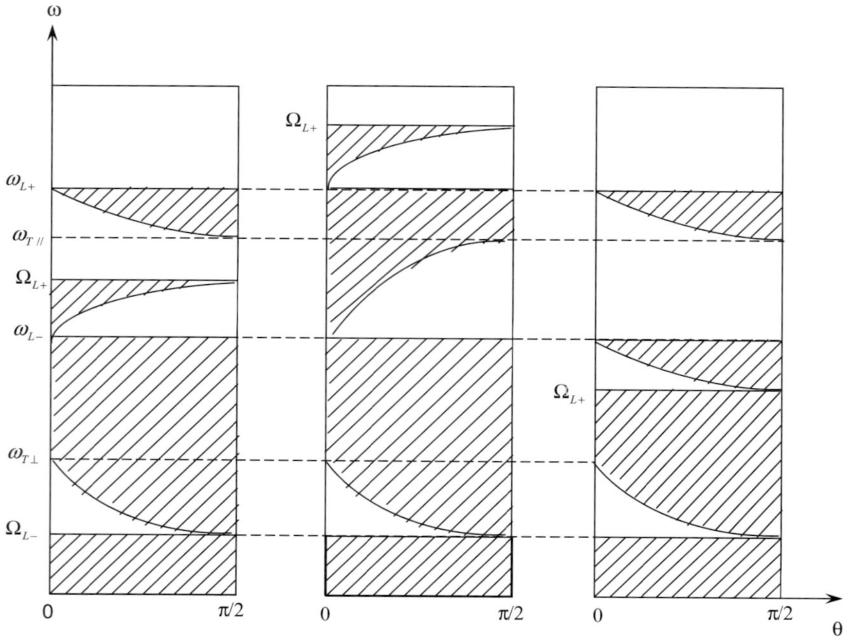

Fig. 2.6 Different mutual positions of the transparency and absorption (shaded) regions for the wave given by Eq. (2.24), plotted as functions of the direction of propagation.

3
Radio Waves and Polaritons in the Presence of an External Static Magnetic Field

3.1
Dielectric Permittivity Tensor at an Arbitrary Orientation of the Magnetic Field with Respect to the Crystal Axis

When an external static magnetic field is applied in a direction not parallel to the crystal axis, propagation of the electromagnetic waves is accompanied by some interesting effects which do not occur in the case of isotropic (cubic) crystals. In this chapter we shall investigate, for example, the shifts in cyclotron and plasma resonances due to their transformation into a coupled cyclotron–plasma resonance and the oscillations of the rotation angle of the polarization ellipse and ellipticity due to the variations of wave frequency, external magnetic field, crystal thickness, and so forth.

As in previous chapters, we shall restrict ourselves to the case of the Maxwellian distribution of the free carriers with ellipsoidal constant-energy surfaces. Moreover, it will be assumed that the frequencies of the considered waves, as well as the cyclotron and plasma frequencies, are much higher than the collision frequency and, consequently, the collisions can be neglected. Before finding the dielectric permittivity tensor of a uniaxial nonpolar crystal with electronic plasma in the presence of a constant magnetic field, let us consider a free carrier with charge "e", whose dynamics is described by the Hamiltonian

$$H = H_0 + H', \qquad H_0 = \frac{\mu_{ij} p_i p_j}{2m}, \tag{3.1}$$

$$H' = -\frac{e}{2m}\mu_{ij}\left(A_i p_j + p_j A_i\right), \tag{3.2}$$

where

$$\boldsymbol{p} = -\mathrm{i}\hbar\nabla - e\boldsymbol{A}_0, \tag{3.3}$$

$\boldsymbol{A}_0$ is the vector potential of the constant magnetic field $\boldsymbol{B}_0 = \text{rot}\,\boldsymbol{A}_0$,

Radiowaves and Polaritons in Anisotropic Media. Roland H. Tarkhanyan and Nikolaos K. Uzunoglu

ISBN: 3-527-40615-8

$$\boldsymbol{A} = \frac{1}{\mathrm{i}\omega}\boldsymbol{E}_0 \exp[\mathrm{i}(\boldsymbol{kr} - \omega t)] \tag{3.4}$$

is the vector potential of the electromagnetic wave.

Let us choose $\boldsymbol{A}_0$ in such a way that for the unperturbed Hamiltonian H_0 we obtain the Schrodinger equation for a harmonic oscillator: $\boldsymbol{A}_0 = \{0, \boldsymbol{B}_0 x, 0\}$. Then $\boldsymbol{B}_0$ is parallel to the z-axis of a Cartesian coordinate system; we assume that the crystal axis lies in the xz-plane and makes an arbitrary angle φ with the z-axis. In this coordinate system the dimensionless tensor of the reciprocal effective mass has the form [28]

$$\boldsymbol{\mu} = \begin{pmatrix} \mu_\perp \cos^2\varphi + \mu_{||}\sin^2\varphi & 0 & (\mu_{||} - \mu_\perp)\sin\varphi\cos\varphi \\ 0 & \mu_\perp & 0 \\ (\mu_{||} - \mu_\perp)\sin\varphi\cos\varphi & 0 & \mu_\perp \sin^2\varphi + \mu_{||}\cos^2\varphi \end{pmatrix} \tag{3.5}$$

The eigenvalues and eigenfunctions of the operator H_0 are given by

$$\varepsilon_n = \hbar\Omega\left(n + \frac{1}{2}\right) + \frac{\mu_{||}\mu_\perp}{\mu_{xx}}\frac{p_z^2}{2m} \tag{3.6}$$

$$\psi_a(\boldsymbol{r}) = \exp\left[\frac{\mathrm{i}}{\hbar}\left(p_y y + p_z z\right)\right]\Phi_n(x - x_a), \tag{3.7}$$

where

$$\Omega = \frac{eB_0}{m}\sqrt{\mu_{xx}\mu_{yy} - \mu_{xy}^2} \tag{3.8}$$

is the cyclotron frequency, $n = 0, 1, 2, \ldots$, a is the set of quantum numbers n, p_y, p_z, which characterize the unperturbed states,

$$x_a = -\frac{R^2 p_y}{\hbar}, \quad R^2 = \frac{\hbar}{eB_0} \tag{3.9}$$

and $\Phi_n(X)$ are the wave functions of the harmonic oscillator, normalized in a unit volume element [42]. For simplicity, we neglect the Zeeman splitting resulting from the spin quantization and only consider the Landau quantization, that is, we assume that the magnetic field is strong and therefore the Landau quantization dominates.

The current density induced in the medium is given by

$$J = eTr(\boldsymbol{\rho v}) = \varepsilon\sum_{\alpha\beta}\rho_{\alpha\beta}\boldsymbol{v}_{\beta\alpha} \tag{3.10}$$

where $\boldsymbol{v}_{\beta a}$ are the matrix elements of the velocity operator $\boldsymbol{v} = \boldsymbol{\mu p}/\mathrm{m}$ and $\boldsymbol{\rho}$ is the one-particle density matrix. The transport equation for $\boldsymbol{\rho}$ has the standard form

$$\mathrm{i}\hbar\partial\boldsymbol{\rho}/\partial t = H\boldsymbol{\rho} - \boldsymbol{\rho}H. \tag{3.11}$$

The linearized solution of Eq. (3.11) can be written as

$$\boldsymbol{\rho} = \boldsymbol{\rho}_0 + \boldsymbol{\rho}' \tag{3.11a}$$

where $\boldsymbol{\rho}_0 = \exp\left(\frac{\varsigma - H_0}{T}\right)$ is the density matrix for the ensemble of particles with Maxwellian equilibrium distribution function and $\boldsymbol{\rho}'$ is the first-order electric field dependent perturbation. Using Eqs. (3.1)–(3.4), (3.7) and (3.10) and the phenomenological expression for the current density $\boldsymbol{J} = \boldsymbol{\sigma E}$, we obtain the conductivity tensor in the form

$$\sigma_{ij}(\omega, \boldsymbol{k}) = \frac{\mathrm{i}Ne^2}{m\omega}\mu_{ij} + \frac{\mathrm{i}e^2}{4\omega m^2}\mu_{is}\mu_{jl}\sum_{a,\beta}\frac{f_{a\beta}I^l_{a\beta}\left(I^s_{\beta a}\right)^*}{\varepsilon_{a\beta} - \hbar\omega}, \tag{3.12}$$

where

$$I^l_{a\beta} = \left[p_l\exp(i\,\boldsymbol{kr}) + \exp(\mathrm{i}\,\boldsymbol{kr})\,p_l\right]_{a\beta}, \tag{3.13}$$

$f_{a\beta} = f_a - f_\beta$, $\varepsilon_{a\beta} = \varepsilon_a - \varepsilon_\beta$, where f_a is the Maxwellian distribution function, normalized by the condition $\sum\limits_a f_a = N$, and N is the density of the carriers.

Evaluating Eq. (3.12) in the absence of spatial dispersion, for the complex dielectric permittivity tensor $\varepsilon_{ij} = \varepsilon^\infty_{ij} + \frac{\mathrm{i}}{\varepsilon_0\omega}\sigma_{ij}$ we obtain

$$\varepsilon_{ij} = \begin{pmatrix} \varepsilon^\infty_{xx} - a\mu_{xx} & -\mathrm{i}ab\sqrt{\mu_\perp\mu_{xx}} & \varepsilon^\infty_{xz} - a\mu_{xz} \\ \mathrm{i}ab\sqrt{\mu_\perp\mu_{xx}} & \varepsilon^\infty_\perp - a\mu_\perp & \mathrm{i}ab\mu_{xz}\sqrt{\mu_\perp\mu^{-1}_{xx}} \\ \varepsilon^\infty_{xz} - a\mu_{xz} & -\mathrm{i}ab\mu_{xz}\sqrt{\mu_\perp\mu^{-1}_{xx}} & \varepsilon^\infty_{zz} - a\mu_{zz} + ab^2\mu_{||}\mu_\perp\mu^{-1}_{xx} \end{pmatrix} \tag{3.14}$$

where

$$a = \frac{\omega_0^2}{\omega^2 - \Omega^2}, \qquad b = \frac{\Omega}{\omega} \tag{3.15}$$

and ω_0 is given by Eq. (1.23). We assume that the principal axes of the tensors ε^∞_{ij} and μ_{ij} coincide; then ε^∞_{ij} can be written in terms of the principal values $\varepsilon^\infty_{||}$ and $\varepsilon^\infty_\perp$:

$$\varepsilon_{ij}^{\infty} = \begin{pmatrix} \varepsilon_{\perp}^{\infty}\cos^2\varphi + \varepsilon_{\|}^{\infty}\sin^2\varphi & 0 & \left(\varepsilon_{\|}^{\infty} - \varepsilon_{\perp}^{\infty}\right)\sin\varphi\cos\varphi \\ 0 & \varepsilon_{\perp}^{\infty} & 0 \\ \left(\varepsilon_{\|}^{\infty} - \varepsilon_{\perp}^{\infty}\right)\sin\varphi\cos\varphi & 0 & \varepsilon_{\perp}^{\infty}\sin^2\varphi + \varepsilon_{\|}^{\infty}\cos^2\varphi \end{pmatrix} \tag{3.16}$$

We have to note that for an isotropic (cubic) crystal, as well as for a uniaxial crystal in the case when the magnetic field $\boldsymbol{B}_0$ is parallel or perpendicular to the crystal axis, the components $\varepsilon_{xz} = \varepsilon_{zx}$ and $\varepsilon_{zy} = \varepsilon_{yz}^{*}$ vanish, and ε_{zz} in Eq. (3.14) has no magnetic field dependence. Unlike this, in a uniaxial crystal both ε_{xz} and ε_{yz} have nonzero values, and ε_{zz} depends on magnetic field for $0 < \varphi < \pi/2$. Moreover, the cyclotron frequency Ω becomes a function of the angle φ between $\boldsymbol{B}_0$ and the crystal axis. Summarizing, the indicated circumstances caused a number of new effects, which are absent in isotropic materials and which are the subjects of the following sections.

3.2
Propagation of Electromagnetic Waves in Uniaxial Nonpolar Semiconductors Along the Magnetic Field $\boldsymbol{B}_0$

Using the general dispersion relation (1.37), we conclude that in an unbounded medium described by the tensor (3.14) two distinct extraordinary waves can travel whose respective refractive indices are given by solutions of the following biquadratic equation:

$$An^4 - Dn^2 + C = 0 \tag{3.17}$$

where

$$A = \varepsilon_{ij}s_i s_j, \quad D = \left(\varepsilon_{ij}\varepsilon_{kk} - \varepsilon_{ik}\varepsilon_{kj}\right)s_i s_j, \quad C = \det(\varepsilon_{ij}). \tag{3.18}$$

In this section we restrict ourselves to the propagation of the waves parallel to $\boldsymbol{B}_0$. Then $s_x = s_y = 0,\ s_z = 1$, and Eq. (3.17) gives

$$n_{\pm}^2 = \frac{1}{2\varepsilon_{zz}}\left\{E_{xx} + E_{yy} \pm \left[\left(E_{xx} - E_{yy}\right)^2 + 4/E_{xy}/^2\right]^{1/2}\right\} \tag{3.19}$$

where E_{ij} is the minor corresponding to the element ε_{ij}:

$$E_{xx} = \varepsilon_{yy}\varepsilon_{zz} - /\varepsilon_{yz}/^2 \qquad E_{yy} = \varepsilon_{xx}\varepsilon_{zz} - \varepsilon_{yz}^2 \tag{3.20}$$

etc. Let us consider the cases $\varphi = 0, \pi/2$ and $\varphi \neq 0, \pi/2$ separately. In the first two cases, a solution of the dispersion equation

$$\varepsilon_{zz} = 0 \tag{3.21}$$

can be separated, which corresponds to the longitudinal wave with a discrete frequency $\omega = \omega_{0\parallel}$ for $\varphi = 0$ and $\omega = \omega_{0\perp}$ for $\varphi = \pi/2$. It is evident that they are plasma oscillations which travel in the directions parallel and perpendicular to the crystal axis [see Eqs. (1.30) and (1.31)]. There are also two transverse waves with circular (for $\varphi = 0$) and elliptical (for $\varphi = \pi/2$) polarizations. The dispersion relations for circularly polarized ($E_y/E_x = \pm \mathrm{i}$) waves are

$$n_{\pm}^2 = \varepsilon_{\perp}^{\infty}\left[1 - \frac{\omega_{0\perp}^2}{\omega(\omega \pm \Omega)}\right] \quad \text{where} \quad \Omega = \frac{eB_0}{m}\mu_{\perp}. \tag{3.22}$$

The (–) wave can be propagated in the frequency regions

$$\omega < \Omega \quad \text{and} \quad \omega > \frac{\Omega}{2}\left(1 + \sqrt{1 + 4\frac{\omega_{0\perp}^2}{\Omega^2}}\right). \tag{3.23}$$

The cyclotron resonance occurs at the boundary of the first region. In this region n_-^2 reaches a minimum value at $\omega = \Omega/2$ which is given by

$$n_-^2 = \varepsilon_{\perp}^{\infty}\left(1 + 4\frac{\omega_{0\perp}^2}{\Omega^2}\right). \tag{3.24}$$

When the magnetic field is increased, this minimum decreases, which corresponds to an increase in the phase-velocity maximum:

$$u_{\max} = \frac{c}{\sqrt{\varepsilon_{\perp}^{\infty}\left(1 + 4\omega_{0\perp}^2/\Omega^2\right)}}. \tag{3.25}$$

At the boundary of the second region (3.23) we find that $n_- = 0$ that is, a cut-off occurs. In this region, the phase velocity decreases with increasing frequency, but is always greater than $u_{\min} = c(\varepsilon_{\perp}^{\infty})^{-1/2}$.

The (+) wave can only be propagated in the region

$$\omega > \frac{\Omega}{2}\left(\sqrt{1 + 4\omega_{0\perp}^2/\Omega^2} - 1\right); \tag{3.26}$$

the latter being its cut-off frequency. At $\omega = \Omega$ this wave does not exhibit a resonance and it can only be propagated if

$$\Omega > \frac{\omega_{0\perp}}{\sqrt{2}}. \tag{3.27}$$

The dispersion relations for the elliptically polarized transverse waves are given by

$$n_{\pm}^2 = \frac{1}{2}\left\{\varepsilon_{\|}^{\infty} + \varepsilon_{\perp}^{\infty} - \frac{(\mu_{\|}+\mu_{\perp})\omega_0^2}{\omega^2-\Omega^2} \pm \left[\left(\varepsilon_{\|}^{\infty} - \varepsilon_{\perp}^{\infty} - \frac{(\mu_{\|}-\mu_{\perp})\omega_0^2}{\omega^2-\Omega^2}\right)^2 + 4\frac{\mu_{\|}\mu_{\perp}\Omega^2\omega_0^4}{\omega^2\left(\omega^2-\Omega^2\right)^2}\right]^{1/2}\right\} \tag{3.28}$$

where

$$\Omega = \frac{eB_0}{m}\sqrt{\mu_{\|}\mu_{\perp}} \tag{3.29}$$

The cyclotron resonance is only possible for the (+) wave. The refractive index for the (–) wave at $\omega = \Omega$ is given by

$$n_{-}^2 = \left(\mu_{\|}\varepsilon_{\perp}^{\infty} + \mu_{\perp}\varepsilon_{\|}^{\infty} - \frac{\omega_0^2}{\Omega^2}\mu_{\|}\mu_{\perp}\right)\left(\mu_{\|}+\mu_{\perp}\right)^{-1}. \tag{3.30}$$

It follows that the wave can only be propagated if

$$\Omega > \frac{\omega_{0\perp}}{\sqrt{1+\varepsilon_{\|}^{\infty}\mu_{\perp}/\varepsilon_{\perp}^{\infty}\mu_{\|}}}. \tag{3.31}$$

The cut-off frequencies of the waves described by Eq. (3.28) are given by

$$\omega_{\pm}^2 = \frac{1}{2}\left\{\Omega^2 + \omega_{0\|}^2 + \omega_{0\perp}^2 \pm \left[\left(\Omega^2 + \omega_{0\|} + \omega_{0\perp}^2\right)^2 - 4\omega_{0\|}^2\omega_{0\perp}^2\right]^{1/2}\right\} \tag{3.32}$$

Let us now inspect the waves for the remaining values of $\varphi \neq 0, \pi/2$. In this case the two waves are neither longitudinal nor transverse, and are elliptically polarized. We shall write the electric field vector as a sum of two vectors:

$$\mathbf{E} = \mathbf{E}_1 + \mathbf{E}_2 \qquad \mathbf{E}_1 = \{1, 0, \lambda\}E_x \qquad \mathbf{E}_2 = \{0, \chi, 0\}E_x \tag{3.33}$$

One of these lies in the xz-plane; the other is parallel to the y-axis and is shifted in phase by $\pi/2$ with respect to the first vector. These two vectors represent the semi-axes of an ellipse described by the end point of the vector $\mathbf{E}$. Using Eq. (1.35) we can show that, for $s_z = 1$

$$\chi_{\pm} = \mathrm{i}(\eta \pm \sqrt{1+\eta^2}) \tag{3.34}$$

$$\lambda_{\pm} = -\frac{\varepsilon_{xz} + \chi_{\pm}\varepsilon_{zy}}{\varepsilon_{zz}} \tag{3.35}$$

$$\eta = \mathrm{i}\frac{E_{yy} - E_{xx}}{2E_{yx}} \tag{3.36}$$

In the case of a nonabsorbing medium η is real. For $\varphi = 0$ as well as for an isotropic medium, $\eta = 0$.

The transverse component $\boldsymbol{E}_{\perp} = \{1, \chi, 0\} E_x$ is also elliptically polarized, and the principal axes of the ellipse are identical with the coordinate axes x and y. As usual, we shall call the polarization right-handed (+) or left-handed (–) depending on the sign of the imaginary part of χ. The phase shifts of the components E_x and E_y in (+) and (–) polarized waves are equal to $\mp\pi/2$ respectively. The polarization in question refers to the transverse component of the electric field $\boldsymbol{E}_{\perp}$. The ellipticity (the ratio of the minor axis to the major axis) is, for both waves, equal to

$$\xi = \sqrt{1+\eta^2} - /\eta/ \tag{3.37}$$

The major (minor) axes of the polarization ellipses of two waves are mutually orthogonal. The major axis of the polarization ellipse of a right-handed polarized wave is parallel to the y-axis if $\eta > 0$ and to the x-axis if $\eta < 0$. For $\eta = 0$, the ellipse becomes a circle whereas at $/\eta/\rightarrow\infty$ it reduces to a straight line. Circular polarization can occur, irrespective of the magnetic field strength, only if $\varphi = 0$ (and also in the isotropic case). For $\varphi \neq 0$ the waves traveling along the magnetic field can have a circular polarization only for a specific value of the field, which is determined from the condition $\eta = 0$ For example, at $\varphi = \pi/2$ this value of the field is given by

$$\Omega^2 = \omega^2 - \omega_0^2 \frac{\mu_{\|} - \mu_{\perp}}{\varepsilon_{\|}^{\infty} - \varepsilon_{\perp}^{\infty}} \tag{3.38}$$

In the case of circular polarization the refractive indices given by Eq. (3.19) are transformed to the form

$$n_{\pm}^2 = \varepsilon_{xx} \mp \mathrm{i}\varepsilon_{xy} \tag{3.39}$$

For $\varphi = 0$ this equation is identical to Eq. (3.22).

Now let us find the resonance frequencies, at which the wave becomes purely longitudinal. Using Eq. (3.19), we obtain

$$\omega^2 = \frac{1}{2}\left\{\Omega^2 + \frac{\mu_{zz}}{\varepsilon_{zz}^{\infty}}\omega_0^2 \pm \left[\left(\Omega^2 + \frac{\mu_{zz}}{\varepsilon_{zz}^{\infty}}\omega_0^2\right)^2 - 4\frac{\mu_{\perp}\mu_{\|}}{\mu_{xx}\varepsilon_{zz}^{\infty}}\omega_0^2\Omega^2\right]^{1/2}\right\} \tag{3.40}$$

Note that for $\varphi = 0$ and $\varphi = \pi/2$ these frequencies are identical with the frequencies of the longitudinal plasma vibrations $\omega_{0\|}$ and $\omega_{0\perp}$, respectively, and with the cyclotron frequency. Thus, for $\varphi \neq 0$, $\pi/2$, the cyclotron and plasma frequencies are shifted; these shifted resonances defined by Eq. (3.40) can be called cyclotron–-plasma resonances. When the magnetic field is increased, both resonance frequencies increase monotonically.

The appearance of the cyclotron–plasma resonances is one of the main features that distinguish the propagation of the waves parallel to $\boldsymbol{B}_0$ in uniaxial semiconductors from that in isotropic (cubic) semiconductors. Moreover, in uniaxial crystals the longitudinal and transverse waves separate only for $\varphi = 0$ and $\varphi = \pi/2$, whereas in all the other cases the waves are neither longitudinal nor transverse. Secondly, the waves in uniaxial crystals are elliptically polarized with the exception of the case $\varphi = 0$, which corresponds to the magnetic field parallel to the crystal axis; in this case, the polarization is circular. And, finally, the refractive indices and the ellipticity of the waves depend essentially on the angle between the magnetic field and the crystal axis.

3.3
Influence of Crystal Anisotropy on the Faraday Magnetooptical Effect

Let us consider a plane-polarized wave

$$\boldsymbol{E}_0 = \left\{E_{0x}, E_{0y}, 0\right\} \exp\left[\mathrm{i}\omega\left(\frac{z}{c} - t\right)\right] \tag{3.41}$$

incident normally from the vacuum on the surface of a uniaxial semiconductor placed in a static magnetic field which is perpendicular to the surface and whose angle with the crystal axis is φ. In the medium, this wave separates into two elliptically polarized waves, which are then propagated independently with refractive indices $n_\pm$ defined by Eq. (3.19). To discuss the configuration of the total electric field

$$\boldsymbol{E} = \left(\boldsymbol{E}_+ \mathrm{e}^{\mathrm{i}\omega n_+ z/c} + \boldsymbol{E}_- \mathrm{e}^{\mathrm{i}\omega n_- z/c}\right) \mathrm{e}^{-\mathrm{i}\omega t} \tag{3.42}$$

where $\boldsymbol{E}_\pm$ are **the** amplitudes of the right- and left-handed polarized waves, it is sufficient to determine the ratio E_y/E_x for an arbitrary value of z. By definition, at the crystal surface $z = 0$ the boundary conditions for the tangential components of the fields give

$$E_{+x} + E_{-x} = E_{0x} \qquad E_{+y} + E_{-y} = E_{0y} \tag{3.43}$$

Using Eqs. (3.33) and (3.43), for the transverse components of the vectors $E_\pm$ we obtain

$$\begin{pmatrix} E_{+x} \\ E_{+y} \end{pmatrix} = a \begin{pmatrix} -\chi_- & 1 \\ -1 & \chi_+ \end{pmatrix} \begin{pmatrix} E_{0x} \\ E_{0y} \end{pmatrix} \tag{3.44}$$

$$\begin{pmatrix} E_{-x} \\ E_{-y} \end{pmatrix} = a \begin{pmatrix} \chi_+ & -1 \\ 1 & -\chi_- \end{pmatrix} \begin{pmatrix} E_{0x} \\ E_{0y} \end{pmatrix} \tag{3.45}$$

where $a = (\chi_+ - \chi_-)^{-1}$ and $\chi_\pm$ are defined by Eq. (3.34). Using Eq. (3.42) and rotating the coordinate axes x and y around the z-axis, in the coordinate system x', y', z, where x' is parallel to the polarization plane of the incident wave, the following equation is easily obtained:

$$\frac{E_{y'}}{E_{z'}} = \frac{1 - i\eta \sin 2\beta}{i\eta \cos 2\beta - (1+\eta)^{1/2} \cot \kappa} \tag{3.46}$$

where

$$\kappa = \frac{\omega}{2c} z(n_+ - n_-), \tag{3.47}$$

η is given by Eq. (3.36) and $\tan \beta = E_{0y}/E_{0x}$.

Equation (3.46) implies that the resulting wave is also elliptically polarized, and that the difference between the phase velocities of the component waves leads to the Faraday rotation of the principal axes of the resulting wave polarization ellipse.

Let us now determine the rotation angle in the frequency region in which both waves exist ($n_\pm^2 > 0$). Writing Eq. (3.46) in the form $E_y/E_x = \exp(i\nu)\tan\rho$ we can calculate the rotation angle ψ for the major axis of the resulting wave polarization ellipse with respect to the polarization plane of the incident wave. The angle of rotation corresponding to a distance z covered by the wave is given by

$$\tan 2\psi = \cos \nu \tan 2\rho = \frac{\eta^2 \sin^2 \kappa \sin 4\beta + (1+\eta^{2)})^{1/2} \sin 2\kappa}{\cos 2\kappa + \eta^2 \left[\cos^2 \kappa + \sin^2 \kappa \cos 4\beta\right]} \tag{3.48}$$

The ellipticity ξ is determined by the equation

$$\frac{2\xi}{1+\xi^2} = \sin \nu \sin 2\rho = \frac{\eta}{1+\eta^2}\left[2\sin^2 \kappa \cos 2\beta - (1+\eta^2)^{1/2} \sin 2\kappa \sin 2\beta\right] \tag{3.49}$$

If $\mathbf{E}_0$ is parallel to one of the coordinate axes ($\beta = l\pi/2,\ l = 0, \pm 1, \pm 2, \ldots$) Eqs. (3.48) and (3.49) simplify to

$$\tan 2\psi = \frac{\sqrt{1+\eta^2} \sin 2\kappa}{\eta^2 + \cos 2\kappa}, \tag{3.50}$$

$$\frac{\xi}{1+\xi^2} = (-1)^l \frac{\eta \sin^2 \kappa}{1+\eta^2}. \tag{3.51}$$

In another particular case, when $\beta = \left(l + \frac{1}{2}\right)\pi/2$, we obtain

$$\tan 2\psi = \frac{\tan 2\kappa}{\sqrt{1+\eta^2}} \tag{3.52}$$

$$\frac{2\xi}{1+\xi^2} = (-1)^{l+1}\frac{\eta \sin 2\kappa}{\sqrt{1+\eta^2}} \tag{3.53}$$

Equations (3.46)–(3.53) provide a complete solution of the problem of the Faraday effect for an arbitrary orientation of the crystal axis with respect to the static magnetic field. In particular, for $\varphi = 0$ we find $\eta = 0$ which yields the well-known result that the wave passing a distance z in an isotropic medium remains linearly polarized and its polarization plane is turned through an angle $\psi = \kappa$.

3.4
Oscillations of the Rotation Angle and the Ellipticity

It follows from Eqs. (3.48)–(3.53) that for $\varphi \neq 0$ the rotation angle ψ and the ellipticity ξ can oscillate when the frequency ω, the magnetic field $B_0 \sim \Omega$ or the distance z are varied. Firstly, such oscillations have been observed in monocrystals of CdS [40] and explained theoretically in Ref. [41]. Let us discuss these oscillations in the case $\varphi = \pi/2$ and $\beta = l\pi/2$ when the crystal axis is parallel to the crystal surface plane and perpendicular to $\boldsymbol{B}_0$, and the electric vector of the incident wave is parallel ($l = 0, \pm 2, \ldots$) or perpendicular ($l = \pm 1, \pm 3, \ldots$) to the crystal axis. Then, using Eqs. (3.14) and (3.36), we obtain

$$\eta = \frac{\omega}{2\Omega}\,\frac{\mu_{\|} - \mu_{\perp} + \left(\varepsilon_{\|}^{\infty} - \varepsilon_{\perp}^{\infty}\right)(\Omega^2 - \omega^2)\omega_0^{-2}}{\sqrt{\mu_{\perp}\mu_{\|}}} \tag{3.54}$$

In the limiting cases $\Omega/\omega \to 0, \infty$ the quantity $|\eta|$ tends to infinity and, as implied by Eqs. (3.50) and (3.51), for $\eta \gg 1$ the following relations hold:

$$\tan 2\psi \approx \eta^{-1} \sin 2\kappa \tag{3.55}$$

$$|\xi| \approx |\eta|^{-1} \sin^2\kappa \tag{3.56}$$

In the range of frequencies and magnetic fields where the condition

$$\omega << \Omega << \omega_0 \tag{3.57}$$

is satisfied, the characteristic oscillation length L defined by the relation

$$2\kappa = \frac{z}{L} \tag{3.58}$$

is equal to

$$L \approx \frac{c\Omega}{\omega\omega_0 / \sqrt{\mu_\perp} - \sqrt{\mu_\parallel} /} \tag{3.59}$$

and can be much shorter than the thickness d of the plate in which the wave is propagated.

If the conditions

$$\frac{\Omega}{\omega} - 1 \equiv \frac{\delta}{\omega} << 1, \tag{3.60a}$$

$$\Omega > \omega > \omega_0 \sqrt{\gamma/2}, \qquad \gamma \equiv \frac{\mu_\parallel + \mu_\perp}{\varepsilon_\parallel^\infty + \varepsilon_\perp^\infty}, \tag{3.60b}$$

are satisfied, for the characteristic length we find

$$L = \frac{\sqrt{2}c\left(\varepsilon_\parallel^\infty + \varepsilon_\perp^\infty\right)^{-1/2}}{\omega / \sqrt{1 - \gamma\omega_0^2/2\Omega^2} - \sqrt{\gamma\omega_0^2/\Omega\delta}} \tag{3.61}$$

In this case the rotation angle and ellipticity at the end of the plate $z = d$ are given by

$$\tan 2\psi = \frac{(\mu_\parallel + \mu_\perp)\sin(d/L)}{2\sqrt{\mu_\parallel\mu_\perp}\left[\cos(d/L) + (\mu_\parallel - \mu_\perp)^2/4\mu_\parallel\mu_\perp\right]}, \tag{3.62}$$

$$\frac{\xi}{1+\xi^2} = \frac{2(-1)^l(\mu_\parallel - \mu_\perp)\sqrt{\mu_\parallel\mu_\perp}}{(\mu_\parallel + \mu_\perp)^2}\sin^2(d/L). \tag{3.63}$$

If δ/ω is a small negative quantity, only one of the waves with different refractive indices can travel in the crystal and the rotation of the polarization ellipse does not occur. For other values of Ω/ω the oscillations of ψ and ξ are described by extremely complicated equations. Since the corresponding experimental data are not available, we shall omit these equations.

3.5 Propagation in the Direction Perpendicular to B_0

Let us now consider the waves for which the wave vector $\boldsymbol{k}$ lies in the xy-plane and makes an angle θ with the x-axis. Then $s_z = 0$ and Eq. (3.17) gives

$$\left(\varepsilon_{xx}\cos^2\theta + \varepsilon_{yy}\sin^2\theta\right)n^4 - \left(E_{xx}\sin^2\theta + E_{yy}\cos^2\theta + E_{zz}\right)n^2 + \det/\varepsilon_{ij}/ = 0 \qquad (3.64)$$

For $\varphi = 0$ and $\varphi = \pi/2$ one of the solutions of this equation corresponds to the ordinary transverse wave with $\boldsymbol{E}||\boldsymbol{B}_0$ and the second describes the extraordinary TM-polarized wave with $\boldsymbol{H}||\boldsymbol{B}_0$. The refractive index of the ordinary wave has no dependence on both the magnetic field and the angle θ and it is given by

$$n^2 = \varepsilon_{||}^{\infty}\left(1 - \frac{\omega_{0||}^2}{\omega^2}\right) \qquad (3.65a)$$

for $\varphi = 0$ and

$$n^2 = \varepsilon_{\perp}^{\infty}\left(1 - \frac{\omega_{0\perp}^2}{\omega^2}\right) \qquad (3.65b)$$

for $\varphi = \pi/2$. The wave can only be propagated at frequencies which exceed the cut-off frequencies $\omega_{0||}$ and $\omega_{0\perp}$, respectively.

The refractive index of the extraordinary wave at $\varphi = 0$ can be written in the form

$$n^2 = \varepsilon_{\perp}^{\infty}\frac{(\omega^2 - \omega_-^2)(\omega^2 - \omega_+^2)}{\omega^2(\omega^2 - \omega_r^2)} \qquad (3.66)$$

where

$$\omega_{\pm}^2 = \omega_{0\perp}^2 + \frac{\Omega^2}{2}\left(1 \pm \sqrt{1 + 4\omega_{0\perp}^2/\Omega^2}\right) \qquad (3.67)$$

correspond to the cut-off frequencies and

$$\omega_r = \sqrt{\Omega^2 + \omega_{0\perp}^2} \qquad (3.68)$$

gives the frequency of the shifted cyclotron–plasma resonance. This wave can be propagated in the following two frequency regions:

$$\omega_- < \omega < \omega_r \qquad \omega > \omega_+ \qquad (3.69)$$

The phase velocity of the wave vanishes at the resonance frequency and tends to infinity at the cut-offs. The group velocity that is given by

$$\boldsymbol{v}_g = d\omega/d\boldsymbol{k} = c\left[\frac{\partial(n\omega)}{\partial\omega}\right]^{-1}\boldsymbol{k}/k \qquad (3.70)$$

vanishes at $\omega = \omega_{+,-}$ and $\omega = \omega_r$. The ratio of the transverse component of the vector $\boldsymbol{E}$ to the longitudinal one is given by

$$\frac{E_\perp}{E_\parallel} = -\mathrm{i}\frac{\omega(\omega^2 - \omega_r^2)}{\Omega\omega_{0\perp}^2} \tag{3.71}$$

which implies that the wave is elliptically polarized and the semiaxes of the polarization ellipse are parallel and perpendicular to the wave vector. At $\omega = \omega_r$ the ellipse reduces to a line and the wave becomes longitudinal; at $\omega = \Omega$ and

$$\omega = \frac{1}{2}\left(\Omega + \sqrt{\Omega^2 + 4\omega_{0\perp}^2}\right) \tag{3.72}$$

the ellipse becomes a circle.

For $\varphi = \pi/2$ the refractive index of the extraordinary wave is given by

$$n^2 = \frac{\varepsilon_\parallel^\infty \varepsilon_\perp^\infty (\omega^2 - \omega_-^2)(\omega^2 - \omega_+^2)}{\left(\varepsilon_\parallel^\infty \cos^2\theta + \varepsilon_\perp^\infty \sin^2\theta\right)\omega^2(\omega^2 - \omega_r^2)} \tag{3.73}$$

where

$$\omega_\pm^2 = \frac{1}{2}\left\{\Omega^2 + \omega_{0\perp}^2 + \omega_{0\parallel}^2 \pm \left[\Omega^2\left(\Omega^2 + 2\omega_{0\parallel}^2 + 2\omega_{0\perp}^2\right) + \left(\omega_{0\parallel}^2 - \omega_{0\perp}^2\right)^2\right]^{1/2}\right\} \tag{3.74}$$

$$\omega_r^2 = \Omega^2 + \frac{\mu_\parallel \cos^2\theta + \mu_\perp \sin^2\theta}{\varepsilon_\parallel^\infty \cos^2\theta + \varepsilon_\perp^\infty \sin^2\theta}\omega_0^2. \tag{3.75}$$

Note that, ω_r unlike $\omega_\pm$ depends on the wave-vector direction with respect to the crystal axis.

The transverse to longitudinal components ratio for the electric field vector is equal to

$$\frac{E_\perp}{E_\parallel} = -\mathrm{i}\sqrt{\frac{\varepsilon_\parallel^\infty}{\varepsilon_\perp^\infty}}\frac{\omega\left(\omega^2 - \Omega^2 - \omega_{0\parallel}^2\right)}{\Omega\omega_{0\parallel}\omega_{0\perp}} \tag{3.76}$$

for $\theta = 0$ and

$$\frac{E_\perp}{E_\parallel} = -\mathrm{i}\sqrt{\frac{\varepsilon_\perp^\infty}{\varepsilon_\parallel^\infty}}\frac{\omega(\omega^2 - \Omega^2 - \omega_{0\perp}^2)}{\Omega\omega_{0\parallel}\omega_{0\perp}} \tag{3.77}$$

for $\theta = \pi/2$. It is interesting that at $\omega = \Omega$, the ellipticities of the wave for $\theta = 0$ and $\theta = \pi/2$ are identical and equal to

$$\xi = \min\left\{\sqrt{\frac{\mu_{||}}{\mu_\perp}}, \sqrt{\frac{\mu_\perp}{\mu_{||}}}\right\}. \tag{3.78}$$

For $\varphi \neq 0, \pi/2$ we shall restrict ourselves to the propagation of the waves in directions parallel ($\theta = 0$) and perpendicular ($\theta = \pi/2$) to the x-axis. Then solutions of Eq. (3.64) correspond to the extraordinary transverse–longitudinal waves with elliptical polarization. For $\theta = 0$, the refractive indices of the waves are given by

$$n_\pm^2 = \frac{1}{2\varepsilon_{xx}}\left\{E_{yy} + E_{zz} \pm \left[\left(E_{yy} - E_{zz}\right)^2 + 4E_{yz}E_{zy}\right]^{1/2}\right\} \tag{3.79}$$

At the frequency

$$\omega = \sqrt{\Omega^2 + \omega_0^2\frac{\mu_{xx}}{\varepsilon_{xx}^\infty}} \tag{3.80}$$

one of the waves has a resonance: $n_- \to \infty$. Note that the resonance frequency given by Eq. (3.80) depends essentially on the angle φ and for $\varphi = \pi/2$ is identical with Eq. (3.68).

Let us now discuss briefly the polarization of the waves described by Eq. (3.79). Writing the electric field vector in the form

$$\mathbf{E} = E_y(\nu, 1, \rho) \tag{3.81}$$

and using the Maxwell equations in the form (1.35), we easily find the following expressions for ν and ρ:

$$\nu_\pm = -\frac{\varepsilon_{xy} + \rho_\pm \varepsilon_{xz}}{\varepsilon_{xx}}, \tag{3.82}$$

$$\rho_\pm = \mathrm{i}\left(\eta_1 \pm \sqrt{1 + \eta_1 2}\right), \tag{3.83}$$

where

$$\eta_1 = \mathrm{i}\frac{E_{zz} - E_{yy}}{2E_{zy}}. \tag{3.84}$$

The transverse components $E_{\perp\pm} = E_y(0, 1, \rho_\pm)$ are also elliptically polarized, and the principal axes of the ellipse are parallel to the coordinate axes y and z.

The ellipticities of the (+) and (–) waves are identical and are given by Eq. (3.31), where η is replaced by η_1.

For $\varphi = \pi/2$, the refractive indices of the extraordinary waves propagating normally to the plane of the crystal axis and a static magnetic field are given by

$$n_{\pm}^2 = \frac{1}{2\varepsilon_{yy}}\left\{E_{xx} + E_{zz} \pm \left[(E_{xx} - E_{zz})^2 + 4E_{xz}^2\right]^{1/2}\right\} \tag{3.85}$$

The resonance frequency is given by Eq. (3.68), where $\Omega = eB_0\mu_\perp/m$ is replaced by $\Omega = eB_0\sqrt{\mu_\perp\mu_{xx}}/m$ Thus, the resonance frequency in this case also depends on the angle φ.

The polarization structure of the waves given by Eq. (3.85) is described by the expression

$$\boldsymbol{E} = E_x(1, \kappa, \lambda) \tag{3.86}$$

where

$$\kappa_{\pm} = \frac{\varepsilon_{xy} + \lambda_{\pm}\varepsilon_{zy}}{\varepsilon_{yy}} \tag{3.87}$$

$$\lambda_{\pm} = \eta_2 \mp \sqrt{1 + \eta_2^2} \tag{3.88}$$

$$\omega_0 = \frac{E_{zz} - E_{xx}}{2E_{zx}} \tag{3.89}$$

Hence, both the waves have an elliptical polarization but, in contrast to the case $\theta = 0$ the transverse components $E_{\perp\pm} = E_x(1.0.\lambda_{\pm})$ are linearly polarized, and are perpendicular to each other. It is evident that the angle $a_{\pm}$ between $\boldsymbol{B}_0$ and $\boldsymbol{E}_{\perp\pm}$ is determined by the equation

$$\cot a_{\pm} = \lambda_{\pm}\,. \tag{3.90}$$

3.6 Voigt Effect in Uniaxial Semiconductors

In this section we intend to investigate the rotation of the polarization plane and the exchange of the ellipticity of the waves propagating in the direction perpendicular to the static magnetic field (Voigt configuration). Both the effects occur due to the presence of the free carriers in a semiconductor. We shall restrict ourselves to the cases $\theta = 0$ ($\boldsymbol{k}$ is parallel to the xz-plane which contains $\boldsymbol{B}_0$ and the crystal axis, case A) and $\theta = \pi/2$ ($\boldsymbol{k}$ is perpendicular to the xz-plane, that is, $\boldsymbol{k}\|oy$, case B).

Case A. Let us assume that a linearly polarized plane wave is incident normally on the crystal surface plane $x = 0$. The crystal separates the wave into two extraordinary waves, which are propagated independently with the respective refractive indices $n_\pm$ defined by Eq. (3.79). The total electric field vector of the waves can be written in the form

$$\mathbf{E}(x,t) = \left(\mathbf{E}_+ e^{i\omega n_+ x/c} + \mathbf{E}_- e^{i\omega n_- x/c}\right) e^{-i\omega t} \tag{3.91}$$

The transverse component of this vector $\mathbf{E}_\perp = (0, E_y, E_z)$ is elliptically polarized, and the difference between the phase velocities of the two component waves leads to the Voigt rotation of the principal axes of the polarization ellipse of the resulting wave. To determine the rotation angle we have to find the ratio E_z/E_y for an arbitrary value of x in the coordinate system y', z', where y' is parallel to the polarization plane of the incident wave $E_0 = \mathbf{E}_\perp(x = 0)$. Using the relations

$$E_{+y} + E_{-y} = E_{0y} \tag{3.92}$$

$$E_{+z} + E_{-z} = E_{0z} \tag{3.93}$$

$$E_{+z} = \rho_\pm E_{\pm y} \tag{3.94}$$

$$\rho_+ \rho_- = 1 \tag{3.95}$$

for the components of the amplitude vectors $\mathbf{E}_+$ and $\mathbf{E}_-$ we obtain

$$E_{+x} = \nu_+ E_{+y}, \quad E_{-x} = \nu_- E_{-y} \tag{3.96}$$

$$E_{+y} = \frac{E_{0z} - \rho_- E_{0y}}{\rho_+ - \rho_-}, \quad E_{-y} = \frac{\rho_+ E_{0y} - E_{0z}}{\rho_+ - \rho_-} \tag{3.97}$$

$$E_{+z} = \frac{\rho_+ E_{0z} - E_{0y}}{\rho_+ - \rho_-}, \quad E_{-z} = \frac{E_{0y} - \rho_- E_{0z}}{\rho_+ - \rho_-} \tag{3.98}$$

where $\nu_\pm$ and $\rho_\pm$ are given by Eqs. (3.82) and (3.83).

Now, using Eqs. (3.91), (3.97) and (3.98), after simple calculations we obtain

$$\frac{E_{z'}}{E_{y'}} = \frac{1 - i\eta_1 \sin 2a_1}{\sqrt{1 + \eta_1^2} \cot \chi_1 - i\eta_1 \cos 2a_1} \tag{3.99}$$

where

$$\chi_1 = \frac{\omega}{2c}(n_+ - n_-)x \tag{3.100}$$

$$\tan a_1 = \frac{E_{0z}}{E_{0y}} \tag{3.101}$$

and η_1 is given by Eq. (3.84).

Equation (3.99) leads to the following relations for the rotation angle ψ of the major axis of the polarization ellipse and for the ellipticity ξ:

$$\tan 2\psi = -\frac{\sin\chi_1(\eta_1^2\sin\chi_1\sin 4a_1 + 2\sqrt{1+\eta_1^2}\cos\chi_1)}{\cos 2\chi_1 + \eta_1^2(\cos^2\chi_1 + \sin^2\chi_1\cos 4a_1)} \tag{3.102}$$

$$\frac{\xi}{1+\xi^2} = \frac{\eta_1\sin\chi_1}{1+\eta_1^2}\left(\sqrt{1+\eta_1^2}\cos\chi_1\sin 2a_1 - \sin\chi_1\cos 2a_1\right) \tag{3.103}$$

When the static magnetic field is applied along ($\varphi = 0$) or across ($\varphi = \pi/2$) the crystal axis, $\eta_1 \to \infty$ and then Eqs. (3.102) and (3.103) give

$$\tan 2\psi = -\frac{\sin 4a_1}{\cot^2\chi_1 + \cos 4a_1} \tag{3.104}$$

$$\frac{2\xi}{1+\xi^2} = \sin 2a_1 \sin 2\chi_1 \tag{3.105}$$

According to Eq. (3.105) the wave is linearly polarized, i.e. $\xi = 0$, only if

$$a_1 = \frac{\pi}{2}l, \quad l = 0, \pm 1, \ldots \tag{3.106}$$

or when

$$\frac{\omega}{c}(n_+ - n_-)x = \pi l \tag{3.107}$$

The wave has a circular polarization ($\xi = 1$) only if the following conditions are satisfied:

$$a_1 = \frac{\pi}{2}\left(l + \frac{1}{2}\right), \quad \frac{\omega}{c}(n_+ - n_-)x = \pi\left(l' + \frac{1}{2}\right), \quad l, l' = 0, \pm 1, \ldots \tag{3.108}$$

In all the other cases the wave is elliptically polarized.

Using Eq. (3.104) we conclude that the Voigt rotation vanishes ($\psi = 0$) only if

$$a_1 = \frac{\pi}{4} l \tag{3.109}$$

or $$\frac{\omega}{c}(n_+ - n_-)x = 2\pi l \tag{3.110}$$

Case B. When the wave is propagated normally to the plane which contains $\mathbf{B}_0$ and the crystal axis, the total electric field in the crystal can be written as

$$\mathbf{E}(y, t) = \left(\mathbf{E}_+ e^{i\omega n_+ y/c} + \mathbf{E}_- e^{i\omega n_- y/c}\right) e^{-i\omega t} \tag{3.111}$$

where $n_\pm$ are given by Eq. (3.85). The configuration of the wave is characterized by the ratio of the transverse components of the vector $\mathbf{E}(y, t)$. In the coordinate system x', y, z' in which z' is parallel to the amplitude vector $\mathbf{E}_0$ of the normally incident wave at the crystal surface plane $y = 0$, for the ratio E_x/E_z at the distance y covered by the wave we obtain

$$\frac{E_{x'}}{E_{z'}} = \frac{1 - \eta_2 \tan 2a_2}{\eta_2 + \tan 2a_2 + i\sqrt{1 + \eta_2^2}\cot\chi_2 \cos^{-1}(2a_2)} \tag{3.112}$$

where

$$\tan a_2 = \frac{E_{0x}}{E_{0z}} \tag{3.113}$$

$$\chi_2 = \frac{\omega}{2c}(n_+ - n_-)y \tag{3.114}$$

and η_2 is defined by Eq. (3.89). The rotation angle and the ellipticity are determined by the following expressions:

$$\tan 2\psi = \frac{2\eta_2 \cos 4a_2 + (1 - \eta_2^2)\sin 4a_2}{(1 + \eta_2^2)\cot^2\chi_2 + 2\eta_2 \sin 4a_2 - (1 - \eta_2^2)\cos 4a_2} \tag{3.115}$$

$$\frac{2\xi}{1 + \xi^2} = \frac{\sin 2\chi_2}{\sqrt{1 + \eta_2^2}}(\eta_2 \sin 2a_2 - \cos 2a_2) \tag{3.116}$$

In the case when $\mathbf{E}_0$ is parallel or perpendicular to the magnetic field (i.e. when $a_2 = \pi l/2,\ l = 0, \pm 1, \ldots$), Eqs. (3.115) and (3.116) simplify to

$$\tan 2\psi = \frac{2\eta_2 \sin^2\chi_2}{\eta_2^2 + \cos 2\chi_2} \tag{3.117}$$

$$\frac{2\xi}{1+\xi^2} = (-1)^{l+1} \frac{\sin 2\chi_2}{\sqrt{1+\eta_2^2}} \tag{3.118}$$

We have to note that in the case of an isotropic material $\eta_2 \to \infty$ and then Eqs. (3.117) and (3.118) give $\omega = \omega_{0\perp}$, i.e. the resulting wave is linearly polarized and the Voigt rotation is absent. Unlike that, in a uniaxial semiconductor at $\varphi \neq 0, \pi/2$ the wave is elliptically polarized and the Voigt rotation occurs.

3.7 Influence of the Magnetic Field on Polaritons in Uniaxial Polar Semiconductors

Infrared optics of uniaxial polar semiconductors has been discussed in Chapter 2. In this section we shall investigate the influence of an external static magnetic field on the propagation of the electromagnetic waves coupled with plasma and optical phonon vibrations. We limit ourselves to crystals with one longitudinal and one transverse optical phonon branch. We neglect the influence of the magnetic field on ions, as well as the spatial dispersion and the dissipation processes. We shall see that the application of the magnetic field changes essentially the spectrum of the polaritons. For example, in the case of the Faraday configuration the number of the branches increases from two without the magnetic field to five, and the crystal becomes transparent in the lowest-frequency region. Unlike the isotropic case, the resonance frequencies of the polaritons turn out to be dependent on the magnetic field.

The electric flux density for a uniaxial polar semiconductor in a static magnetic field can be written in the form

$$\boldsymbol{D} = \varepsilon_0 \boldsymbol{E} + \boldsymbol{P} + \frac{\mathrm{i}}{\omega} \boldsymbol{\sigma} \boldsymbol{E} \tag{3.119}$$

where $\boldsymbol{P}$ is the polarization density vector of the crystal lattice given by Eq. (1.69) and $\boldsymbol{\sigma}$ is the conductivity tensor given by Eq. (3.12). Using Eq. (3.119) and our previous notation, for the complex permittivity tensor we obtain an expression which differs from Eq. (3.14) only by the replacement of $\varepsilon_{ij}^{\infty}$ by the tensor $\varepsilon_{\mathrm{L}}(\omega)$ whose principal values $\varepsilon_{\mathrm{L}\|}(\omega)$ and $\varepsilon_{\mathrm{L}\perp}(\omega)$ are given by Eq. (1.77). In the coordinate system in which the z-axis is parallel to $\boldsymbol{B}_0$ and the crystal axis lies in the xz-plane, the components of $\varepsilon_{\mathrm{L}}(\omega)$ can be expressed in terms of the principal values analogously to those for the tensor ε^{∞} given by Eq. (3.16). Then, substituting the expression for ε_{ij} into the general dispersion relation (3.17), we can investigate the properties of the waves propagating in an arbitrary direction with respect to the magnetic field. In the following we shall consider both the Faraday and the Voigt configurations in various cases of the crystal-axis orientation with respect to $\boldsymbol{B}_0$.

3.7.1
Propagation along B_0

The dispersion equation has two solutions corresponding to the extraordinary waves whose refractive indices are given by Eq. (3.19). When the crystal axis is parallel ($\varphi = 0$) or perpendicular ($\varphi = \pi/2$) to $\boldsymbol{B}_0$, these waves split into a longitudinal wave with the dispersion relation (3.21) and two transverse waves, for which

$$n^2 = \varepsilon_{xx} \pm /\varepsilon_{xy}/ \tag{3.120}$$

if $\varphi = 0$ and

$$n^2 = \frac{1}{2}\left\{\varepsilon_{xx} + \varepsilon_{yy} \pm \left[\left(\varepsilon_{xx} - \varepsilon_{yy}\right)^2 + 4/\varepsilon_{xy}/^2\right]^{1/2}\right\} \tag{3.121}$$

if $\varphi = \pi/2$. Equation (3.21) determines two discrete frequencies of the longitudinal vibrations which are independent of the magnetic field and which coincide with those given by Eq. (2.21) for $\varphi = 0$ and Eq. (2.22) for $\varphi = \pi/2$. For $\varphi \neq 0, \pi/2$, Eq. (3.21) gives

$$\varepsilon_{L\perp}(\omega)\sin^2\varphi + \varepsilon_{L\|}(\omega)\cos^2\varphi - \mu_{zz}\frac{\omega_0^2}{\omega^2}\left[1 + \frac{\Omega^2}{\omega^2 - \Omega^2}\left(1 - \frac{\mu_{\|}\mu_{\perp}}{\mu_{xx}\mu_{zz}}\right)\right] = 0 \tag{3.122}$$

which determines four different frequencies for the longitudinal (rot $\boldsymbol{E} = 0$) waves corresponding to the mixed cyclotron–plasma and longitudinal–transverse optical phonon modes. Each of two pairs of these coupled modes can be obtained if we neglect the lattice or the electronic plasma contributions to ε_{zz}. Frequencies of "pure" cyclotron–plasma modes are given by Eq. (3.40), and those for "pure" optical phonon modes by Eq. (1.83), where θ should be replaced by φ. For each of the four frequency values satisfying Eq. (3.122), Eq. (3.19) gives two values of the refractive index, one of which is infinite, i.e. corresponds to the resonance.

In Fig. 3.1, the magnetic field dependence of the resonance frequencies is shown schematically. The points ω_1, ω_2 and ω_3 correspond to the frequencies of the phonon–polariton vibrations in the absence of the magnetic field and are given by solutions of the following bicubic equation:

$$\varepsilon_{Lzz}(\omega) - \mu_{zz}\frac{\omega_0^2}{\omega^2} = 0 \tag{3.123}$$

We see that the frequencies of all the longitudinal branches increase with increasing magnetic field. As $\Omega \to \infty$, the three lower frequencies approach the finite values that are given by solutions of the equation

$$\varepsilon_{Lzz}(\omega) - \left(\mu_{zz} - \frac{\mu_{xz}^2}{\sqrt{\mu_{\perp}\mu_{xx}}}\right)\frac{\omega_0^2}{\omega^2} = 0 \tag{3.124}$$

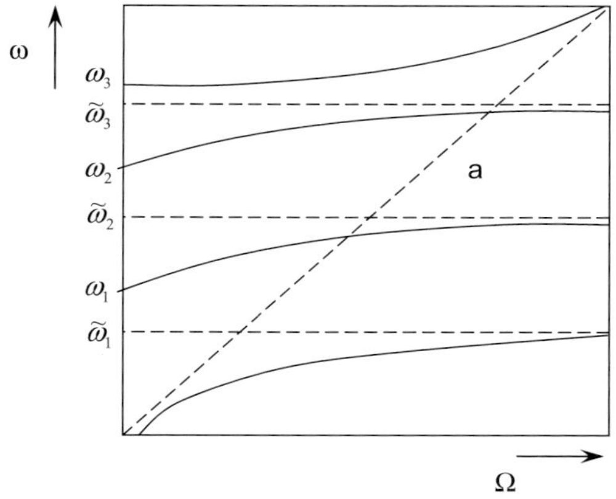

Fig. 3.1 Magnetic field dependence of the resonance frequencies given by solutions of Eq. (3.123).

and the highest frequency approaches the line $\omega = \Omega$ asymptotically.

We have to note that in an isotropic medium the resonance frequencies do not depend on the magnetic field. This dependence, as well as the doubling of the number of the resonances, is the result of the crystal anisotropy and takes place only if $\varphi \neq 0, \pi/2$.

The transverse waves with the dispersion relation (3.120) have circular polarization $(E_y/E_x = \pm \mathrm{i})$ and those described by Eq. (3.121) have elliptical polarization:

$$\frac{E_y}{E_x} = \frac{\mathrm{i}}{2/\varepsilon_{xy}/}\left[\varepsilon_{yy} - \varepsilon_{xx} \pm \sqrt{4/\varepsilon_{xy}/^2 + \left(\varepsilon_{yy} - \varepsilon_{zz}\right)^2}\right] \tag{3.125}$$

Substituting the expressions for ε_{xx} and ε_{xx} at $\varphi = 0$ into Eq. (3.120), we obtain

$$n_{\pm}^2 = \varepsilon_{\perp}^{\infty}\left[\frac{\omega^2 - \omega_{\mathrm{L}\perp}^2}{\omega^2 - \omega_{\mathrm{T}\perp}^2} - \frac{\omega_{0\perp}^2}{\omega(\omega \pm \Omega)}\right] \tag{3.126}$$

The (–) wave exists $(n_{-}^2 > 0)$ in the following three frequency regions:

$$\omega < \min(\Omega, \omega_{\mathrm{T}\perp}), \quad \omega_1^- < \omega < \max(\Omega, \omega_{\mathrm{T}\perp}), \quad \omega < \omega_2^- \tag{3.127}$$

Here $\omega_{\mathrm{T}\perp}$ and Ω correspond to the resonances and $\omega_{1,2}^-$ to the cut-off frequencies, which coincide with the positive real solutions of the equation

$$\omega^4 - \Omega\omega^3 - (\omega_{\mathrm{L}\perp}^2 + \omega_{0\perp}^2)\omega^2 + \omega_{\mathrm{L}\perp}^2\omega + \Omega\omega_{\mathrm{T}\perp}^2\omega_{0\perp}^2 = 0 \tag{3.128}$$

Both the cut-off frequencies increase with increasing magnetic field (Fig. 3.2).

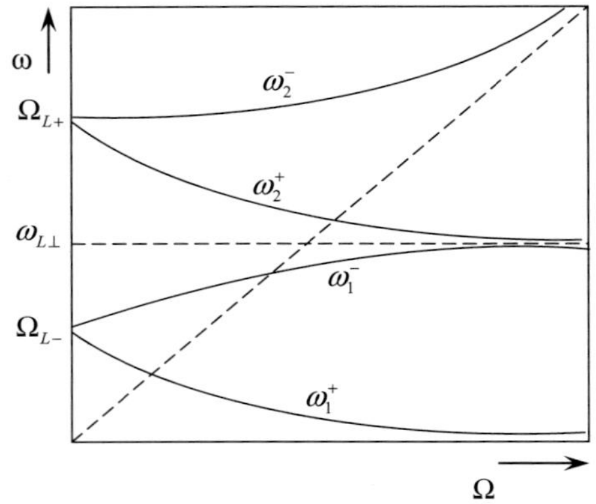

Fig. 3.2 Magnetic field dependence of the cut-off frequencies for the transverse waves described by Eq. (3.126). $\Omega_{L\pm}$ are given by Eq. (2.22).

We note that in a nonpolar crystal the (–) wave has only one resonance at the cyclotron frequency, one cut-off and two regions of existence. In the presence of the lattice vibrations, the same situation occurs only if $\Omega = \omega_{T\perp}$. Thus, the number of both the cut-off and resonance frequencies is doubled and an additional region of propagation appears when the optical phonons are taken into account.

The wave with refractive index n_+ has a resonance at $\omega = \omega_{T\perp}$, two cut-offs at $\omega = \omega_{1,2}^+$ and exists in two frequency regions:

$$\omega_1^+ < \omega < \omega_{T\perp}, \quad \omega > \omega_2^+ \tag{3.129}$$

The cut-off frequencies are given by the real positive solutions of Eq. (3.128), where Ω is replaced by $-\Omega$. Both ω_1^+ and ω_2^+ decrease with increasing magnetic field (Fig. 3.2).

In Fig. 3.3, $n_\pm^2$ are plotted as functions of ω in the regions of existence, for different cases of relative arrangement of the characteristic frequencies. It is seen that when the frequency changes, regions of transparency and nontransparency follow in turn after one another. In general, there are two frequency regions of nontransparency for the transverse waves when $\Omega > \omega_{T\perp}$ (Fig. 3.3a):

$$\omega_{T\perp} < \omega < \omega_1^- \qquad \Omega < \omega < \omega_2^+ \tag{3.130}$$

and only one when $\Omega > \omega_{T\perp}$ (Fig. 3.3b):

$$\omega_{T\perp} < \omega < \omega_2^+ \tag{3.131}$$

The second region in Eq. (3.130) and the region (3.131) extend towards higher frequencies with increasing density of conduction electrons and narrow with increasing magnetic field. Unlike this, the first region in Eq. (3.130) narrows with increasing density of the free carriers and widens with increasing B_0.

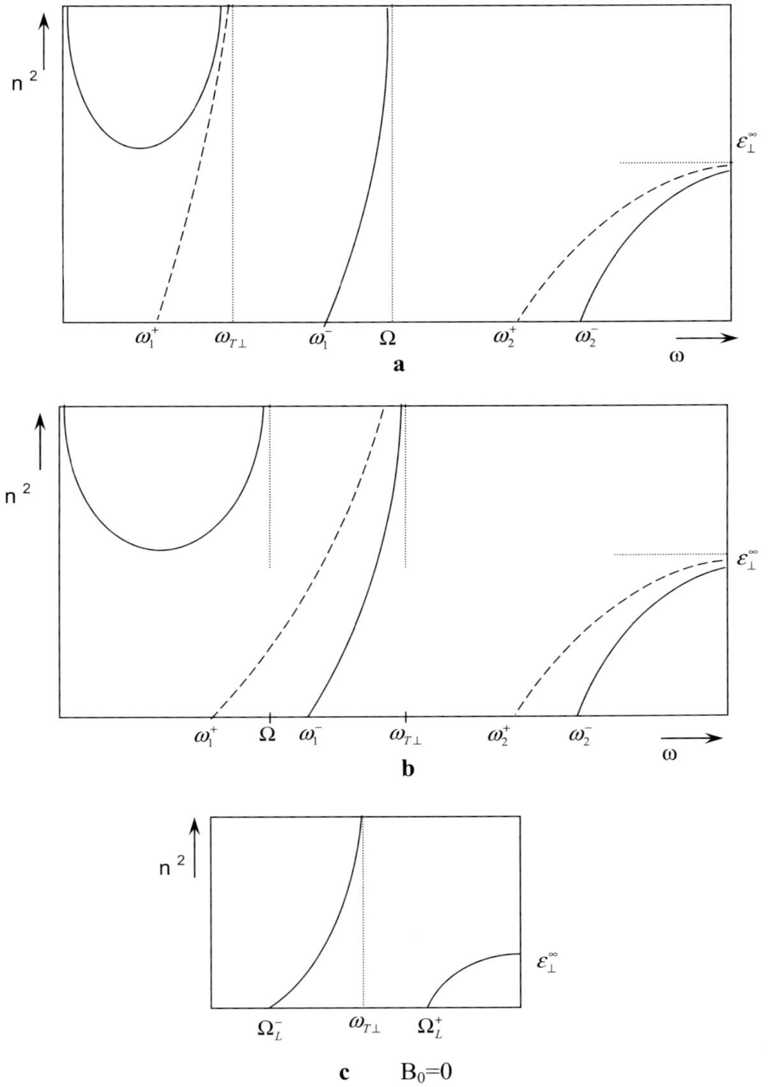

Fig. 3.3 Frequency dependence of the refractive indices for the transverse waves given by Eq. (3.126), in the transparency regions. Solid lines denote (–) waves, dashed lines denote (+) waves.

For comparison, in Fig. 3.3c n^2 versus ω is shown in the existence regions of the transverse polaritons in the absence of the magnetic field. We see that the magnetic field changes essentially the characteristics of the polaritons; particularly, five coupled branches appear instead of two transverse branches, and the crystal becomes transparent in the lowest-frequency region.

3.7.2
Propagation in the Case of the Voigt Configuration

Here we shall only consider the magnetic field parallel to the crystal axis. Then the dispersion equation splits in two. One relation

$$n^2 = \varepsilon_{\|}^{\infty} \left(\frac{\omega^2 - \omega_{\mathrm{L}\|}^2}{\omega^2 - \omega_{\mathrm{T}\|}^2} - \frac{\omega_{0\|}^2}{\omega^2} \right) \tag{3.132}$$

corresponds to the transverse wave with $\boldsymbol{E} \| \boldsymbol{B}_0$ and is identical to Eq. (2.28). The magnetic field has no influence on the propagation of this wave. The second relation

$$n^2 = \varepsilon_{xx} - \frac{/\varepsilon_{xy}/^2}{\varepsilon_{xx}} \tag{3.133}$$

corresponds to the extraordinary wave with $E \perp B_0$. Substituting the expressions for ε_{xx} and ε_{xy} at $\varphi = 0$ we can rewrite Eq. (3.133) in the form

$$n^2 = \varepsilon_{\perp}^{\infty} \frac{(\omega - \omega_1^-)(\omega - \omega_2^-)(\omega - \omega_1^+)(\omega - \omega_2^+)P_4(\omega)}{\omega^2(\omega^2 - \omega_{\mathrm{T}\perp}^2)(\omega^2 - \Omega_+^2)(\omega^2 - \Omega_-^2)} \tag{3.134}$$

where $P_4(\omega)$ is a positive polynomial of the fourth order and

$$\Omega_{\pm}^2 = \frac{1}{2}\left\{ \Omega^2 + \omega_{0\perp}^2 + \omega_{\mathrm{L}\perp}^2 \pm \left[(\Omega^2 + \omega_{0\perp}^2 + \omega_{\mathrm{L}\perp}^2)^2 - 4(\Omega^2\omega_{\mathrm{L}\perp}^2 + \omega_{0\perp}^2\omega_{\mathrm{T}\perp}^2 \right]^{1/2} \right\} \tag{3.135}$$

Thus, the extraordinary wave described by Eq. (3.134) has four cut-off frequencies coinciding with those for the transverse wave propagating along $\boldsymbol{B}_0$; it has three resonances at $\omega = \Omega_{\pm}$ and $\omega = \omega_{\mathrm{T}\perp}$, and exists in four frequency regions. If $\omega_{\mathrm{T}\perp} < \Omega_-$, these regions are

$$\omega_1^- < \omega < \omega_{\mathrm{T}\perp}, \quad \omega_1^+ < \omega < \Omega_-, \quad \omega_2^- < \omega < \Omega_+, \quad \omega > \omega_2^+ \tag{3.136}$$

and if $\Omega_- < \omega_{\mathrm{T}\perp}$ they are interchanged in (3.136). If $\Omega = \omega_{\mathrm{T}\perp}$, the number of the transparent regions diminishes by one. Unlike the case $\boldsymbol{k} \| \boldsymbol{B}_0$, the propagation of a lower-frequency wave with $\omega < \omega_1^-$ is impossible due to the total reflection. In this respect, the situation is analogous to the case $B_0 = 0$, but the edge of the non-transparency region depends on the magnitude of the magnetic field and is shifted towards lower frequencies.

The wave described by Eq. (3.134) is elliptically polarized:

$$\frac{E_y}{E_x} = \frac{\varepsilon_{xx}}{\varepsilon_{yx}} = \mathrm{i}\frac{\omega(\omega^2 - \Omega_+^2)(\omega^2 - \Omega_-^2)}{\Omega\omega_{0\perp}^2(\omega^2 - \omega_{\mathrm{T}\perp}^2)} \tag{3.137}$$

At the resonance frequencies the ellipse transforms into a line; the wave becomes a longitudinal one at $\omega = \Omega_{\pm}$ and a transverse one at $\omega = \omega_{T\perp}$. At the cut-off frequencies and at $\omega = \Omega$ the ellipse transforms into a circle. For a given value of ω the circular polarization takes place only for two values of the magnetic field, for which

$$\Omega = \omega \quad \text{and} \quad \Omega = \omega/\nu/ \tag{3.138}$$

and a linear polarization occurs only if

$$\Omega = \omega\sqrt{\nu} \tag{3.139}$$

where

$$\nu = 1 - \frac{\omega_{0\perp}^2\left(\omega^2 - \omega_{T\perp}^2\right)}{\omega^2\left(\omega^2 - \omega_{L\perp}^2\right)} \tag{3.140}$$

As can be easily seen, Eq. (139) is only satisfied in the frequency regions

$$\Omega_{L-} < \omega < \omega_{L\perp} \quad \text{and} \quad \omega > \Omega_{L+} \tag{3.141}$$

where $\Omega_{L\pm}$ are defined by Eq. (2.22).

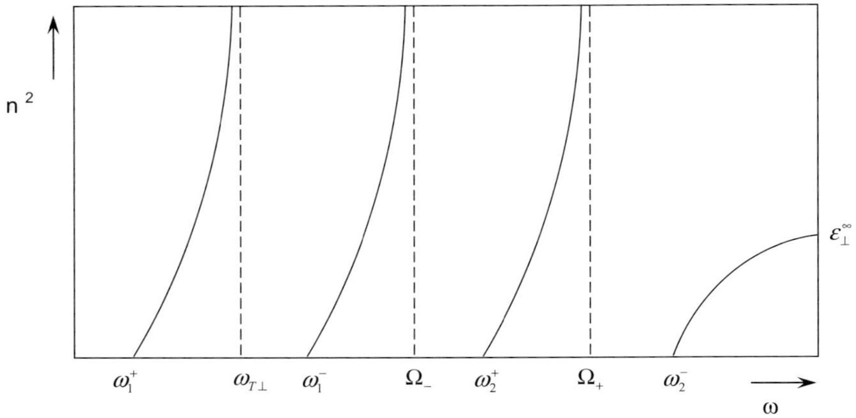

Fig. 3.4 Frequency dependence of n^2 for the extraordinary wave given by Eq. (3.134), in the regions of transparency.

In Fig. 3.4, the function $n^2(\omega)$ given by Eq. (3.134) is shown, in the regions of transparency. As we recall, in the absence of a magnetic field in the direction perpendicular to the crystal axis there can propagate a transverse wave described by Eq. (3.132), two longitudinal phonon–plasmon waves with discrete frequencies

$\omega = \Omega_{L\pm}$ and an ordinary wave described by Eq. (2.23). An external magnetic field applied along the crystal axis replaces the last three waves by the extraordinary wave (3.134), which leads to the considerable modification of the optical properties of the crystal under consideration in the far-infrared region.

4
Reflection of Electromagnetic Waves From the Surface of Uniaxial Semiconductors

4.1
Reflection of s-Polarized Waves From the Surface of a Semi-Infinite Nonpolar Crystal

In this chapter we turn to a consideration of the reflection of plane monochromatic high-frequency ($\omega\tau >> 1$) electromagnetic waves from the surface of a uniaxial crystal with collisionless electronic plasma. As in previous chapters, here we shall restrict ourselves to cases in which it is permissible to neglect the spatial dispersion.

We start from a consideration of various cases of reflection of a so-called s-polarized incident wave, in which the electric field vector is perpendicular to the plane of incidence, containing the wave vector $\boldsymbol{k}_0 = \{k_{0x}, 0, k_{0z}\}$ and the normal to the surface. The crystal is considered as filling the space region $z > 0$; the region $z < 0$ is taken to be a vacuum. The reflected wave is also s-polarized, with refractive index $n = 1$, i.e. $k = \omega/c$. The refracted wave is an ordinary transverse wave with refractive index $n(\omega)$, which depends on the crystal-axis orientation with respect to the reflecting surface. Since there is complete homogeneity in the xy-plane, tangential components of the wave vectors are the same for all three waves and are equal to

$$k_{0x} = \frac{\omega}{c}\sin\psi \equiv k_{||} \tag{4.1}$$

where ψ is the angle of incidence: $\tan\psi = k_{0x}/k_{0z}$. The normal component of the wave vector is equal to $-k_{0z} = -\omega\cos\psi/c$ for the reflected wave and

$$k_{\perp} = \frac{\omega}{c}\sqrt{n^2 - \sin^2\psi} \tag{4.2}$$

for the refracted wave. Here $n^2 = \varepsilon_{\perp}(\omega)$ if the crystal axis $\boldsymbol{C}$ lies in the incident plane (case A) and $n^2 = \varepsilon_{||}(\omega)$ if $\boldsymbol{C}$ is perpendicular to that plane (case B); $\varepsilon_{||,\perp}(\omega)$ are given by Eqs. (1.30) and (1.31), where $\varepsilon^{\mathrm{L}}_{||,\perp} = \varepsilon^{\infty}_{||,\perp}$.

Radiowaves and Polaritons in Anisotropic Media. Roland H. Tarkhanyan and Nikolaos K. Uzunoglu

ISBN: 3-527-40615-8

Using the standard boundary conditions at the reflecting plane $z = 0$ and Eq. (1.33a) for elimination of the tangential components of magnetic field vectors, we can write the continuity conditions in the form

$$E_0 + E_R = E, \quad (E_0 - E_R)\cos\psi = \frac{c}{\omega}k_\perp E \tag{4.3}$$

where E_0, E_R and E are the amplitudes of the incident, reflected and refracted waves, respectively. From Eqs. (4.3), for the coefficient of reflection $R_s = |E_R/E_0|^2$ we obtain the Fresnel equation

$$R_s = \left|\frac{\cos\psi - \sqrt{n^2 - \sin^2\psi}}{\cos\psi + \sqrt{n^2 - \sin^2\psi}}\right|^2 \tag{4.4}$$

The angle of reflection is always equal to the angle of incidence, and the angle of refraction (ψ_1) is given by

$$\sin\psi_1 = \frac{\sin\psi}{n} \tag{4.5a}$$

or

$$\tan\psi_1 = \frac{\sin\psi}{\sqrt{n^2 - \sin^2\psi}} \tag{4.5b}$$

Using Eqs. (4.5a) and (4.5b), we can transform Eq. (4.4) to the form

$$R_s = \left|\frac{\tan\psi - \tan\psi_1}{\tan\psi + \tan\psi_1}\right|^2 = \frac{\sin^2(\psi - \psi_1)}{\sin^2(\psi + \psi_1)} \tag{4.6}$$

As follows from Eqs. (4.5a), (4.5b) and (4.6), only at $n = 1$ does one have $\psi_1 = \psi$ and $R_s = 0$. In the case A, R_s vanishes at $\omega = \omega_2$, where

$$\omega_2 = \omega_{0\perp}\sqrt{\frac{\varepsilon_\perp^\infty}{\varepsilon_\perp^\infty - 1}} \tag{4.7}$$

In the case B, the index $\perp$ in Eq. (4.7) should be replaced by the index $\|$.

As the frequency is varied, regions of transmission and total reflection alternate. In the case A, for $\omega \leq \omega_{0\perp}$ there is no transmission, whatever the angle of incidence. In the frequency region

$$\omega_{0\perp} < \omega < \omega_2 \tag{4.8}$$

transmission occurs only if the condition

$$\sin\psi < \sqrt{\varepsilon_\perp^\infty} \tag{4.9}$$

is fulfilled. For $\omega \geq \omega_2$, there is transmission, whatever the angle ψ. For a given value of $\psi \neq 0$ transmission only occurs in the region $\omega > \omega_1$, where

$$\omega_1 = \frac{\omega_{0\perp}}{\sqrt{1 - (\varepsilon_\perp^\infty)^{-1}\sin^2\psi}} \tag{4.10}$$

With increasing ω in this region, the coefficient of reflection R_s decreases rapidly, vanishes at $\omega = \omega_2$ and then increases monotonically, R_s always being less than 1 (see Fig. 4.1).

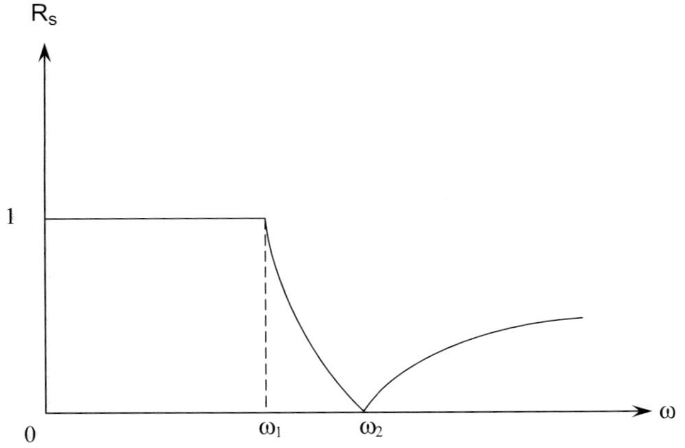

Fig. 4.1 Frequency dependence of the reflection coefficient R_S given by Eq. (4.4).

Note that for a given value of ψ and a known magnitude of the free carrier density, the limit of the regions of transmission and total reflection (ω_1) and the characteristic frequency ω_2 depend only on the transverse components $\varepsilon_\perp^\infty$ and $\mu_\perp$, which provides a possibility of measuring the latter. In the case B, one must replace $\varepsilon_\perp^\infty$ and $\mu_\perp$ by $\varepsilon_\parallel^\infty$ and $\mu_\parallel$ in Eqs. (4.8)–(4.10).

In the case of many-valley cubic crystals and a reflecting surface parallel to the face of the cube, one merely needs to replace $\mu_\perp$ by $\mu = \frac{1}{3}\left(2\mu_\perp + \mu_\parallel\right)$ in the relations above.

4.2 Reflection in the Case of a p-Polarized Incident Wave

Let us now consider the reflection of a so-called p-polarized wave, in which the electric field lies in the plane of incidence. We shall start from the simplest case when the reflecting surface is perpendicular to the crystal axis. Then, the refracted wave is an extraordinary TM-polarized one with the angle of refraction ϑ and refractive index given by Eq. (1.40), which can be rewritten in the form

$$\frac{c^2}{\omega^2}\left[\frac{k_\perp^2}{\varepsilon_\perp(\omega)}+\frac{k_\parallel^2}{\varepsilon_\parallel(\omega)}\right]=1 \tag{4.11}$$

Using Eq. (1.32) and eliminating the tangential components E_{0x}, E_{Rx} and E_x of the electric field vectors, we obtain the following boundary conditions in the plane $z = 0$:

$$H_0+H_R=H,\quad (H_0-H_R)\cos\psi=\frac{ck_\perp}{\omega\varepsilon_\perp(\omega)}H \tag{4.12}$$

These equations, with Eqs. (4.1) and (4.11), lead to the following expression for the coefficient of reflection $R_p = |H_R/H_0|^2$

$$R_p=\left|\frac{\sqrt{\varepsilon_\parallel(\omega)\cos^{-2}\psi-\tan^2\psi}-\sqrt{\varepsilon_\parallel(\omega)\varepsilon_\perp(\omega)}}{\sqrt{\varepsilon_\parallel(\omega)\cos^{-2}\psi-\tan^2\psi}+\sqrt{\varepsilon_\parallel(\omega)\varepsilon_\perp(\omega)}}\right|^2 \tag{4.13}$$

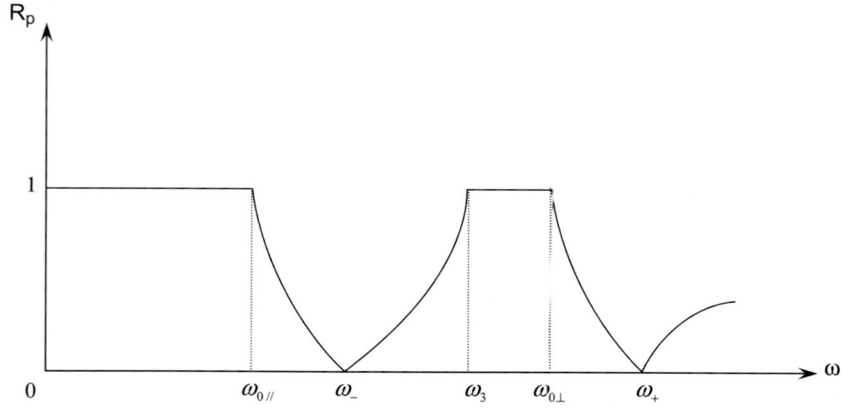

Fig. 4.2 Frequency dependence of the reflection coefficient R_p, in the case when the crystal axis is perpendicular to the reflecting plane [see Eq. (4.19)].

In Fig. 4.2, the frequency dependence of R_p is shown schematically for the case when $\omega_{0\perp} > \omega_{0\parallel}$ and $\sin^2\psi < \varepsilon_\parallel^\infty - \mu_\parallel\mu_\perp^{-1}\varepsilon_\perp^\infty$. Transmission is only possible in two frequency regions:

$$\omega_{0\parallel}<\omega<\omega_3,\quad \omega>\omega_{0\perp} \tag{4.14}$$

where $$\omega_3 = \frac{\omega_{0\|}}{\sqrt{1 - \left(\varepsilon_{\|}^{\infty}\right)^{-1} \sin^2 \psi}} \tag{4.15}$$

In the first region transmission occurs only if

$$\sin \psi < \sqrt{\varepsilon_{\|}^{\infty}} \tag{4.16}$$

Note that, in contrast to the uniaxial crystals, cubic crystals have only one region of transmission ($\omega_{0\|} = \omega_{0\perp}$).

In both regions (4.14), $R_p(\omega)$ initially decreases with increasing ω, vanishes at $\omega = \omega_-$ and $\omega = \omega_+$ and then increases again. Here $\omega_\pm$ are given by solutions of the biquadratic equation

$$(a + b)\omega^4 - \left(\omega_{0\perp}^2 + a\omega_{0\|}^2\right)\omega^2 + \omega_{0\|}^2\omega_{0\perp}^2 = 0 \tag{4.17}$$

where

$$a = 1 - \frac{1}{\varepsilon_{\perp}^{\infty} \cos^2 \psi}, \quad b = \frac{\tan^2 \psi}{\varepsilon_{\|}^{\infty} \varepsilon_{\perp}^{\infty}} \tag{4.17a}$$

At a given value of ω, $R_p(\omega)$ vanishes only if the angle of incidence is equal to the Brewster angle ψ_B, where

$$\tan^2 \psi_B = \frac{\varepsilon_{\|}(\omega)[\varepsilon_{\perp}(\omega) - 1]}{\varepsilon_{\|}(\omega) - 1} \tag{4.18}$$

For the angle of refraction ϑ we obtain

$$\cot^2 \vartheta = \frac{k_{\perp}^2}{k_{\|}^2} = \frac{\varepsilon_{\perp}(\omega)\left[\varepsilon_{\|}(\omega) - \sin^2 \psi\right]}{\varepsilon_{\|}(\omega)\sin^2 \psi} \tag{4.19}$$

Then Eq. (4.13) can be rewritten in the form

$$R_p = \left|\frac{\varepsilon_{\perp}(\omega)\cot \psi - \cot \vartheta}{\varepsilon_{\perp}(\omega)\cot \psi + \cot \vartheta}\right|^2 \tag{4.20}$$

Note that at $\psi = \psi_B$, the angle of refraction is given by

$$\cot \vartheta = \varepsilon_{\perp}(\omega)\cot \psi_B \tag{4.21}$$

In the case when the crystal axis is parallel to the reflecting surface and lies in the plane of incidence, in Eqs. (4.11)–(4.21) $\varepsilon_\perp^\infty$ and $\mu_\perp$ should be replaced by $\varepsilon_\parallel^\infty$ and $\mu_\parallel$, and *vice versa*.

Now let us consider another interesting case when the crystal axis is perpendicular to the plane of incidence. Then the refracted wave is an ordinary one, and $k_\perp$ in Eqs. (4.12) is given by Eq. (4.2), where $n^2 = \varepsilon_\perp(\omega)$. For the coefficient of reflection we obtain

$$R_\mathrm{p} = \left| \frac{\varepsilon_\perp(\omega) - \sqrt{\varepsilon_\perp(\omega) + [\varepsilon_\perp(\omega) - 1]\tan^2\psi}}{\varepsilon_\perp(\omega) + \sqrt{\varepsilon_\perp(\omega) + [\varepsilon_\perp(\omega) - 1]\tan^2\psi}} \right|^2 \tag{4.22}$$

Transmission occurs only if the condition $\sin\psi < \sqrt{\varepsilon_\perp^\infty}$ is satisfied. The transmission region is $\omega > \omega_1$, which is given by Eq. (4.10). In this region R_p vanishes at two different frequencies (see Fig. 4.3), one of which coincides with ω_2 given by Eq. (4.7), while the second is given by

$$\omega_4 = \frac{\omega_{0\perp}}{\sqrt{1 - (\varepsilon_\perp^\infty)^{-1}\tan^2\psi}} \tag{4.23}$$

and exists only if $\tan\psi < \sqrt{\varepsilon_\perp^\infty}$. At the frequency $\omega = \omega_5$, where

$$\omega_5 = \frac{\omega_{0\perp}}{\sqrt{1 - 2(\varepsilon_\perp^\infty)^{-1}\sin^2\psi}} \tag{4.24}$$

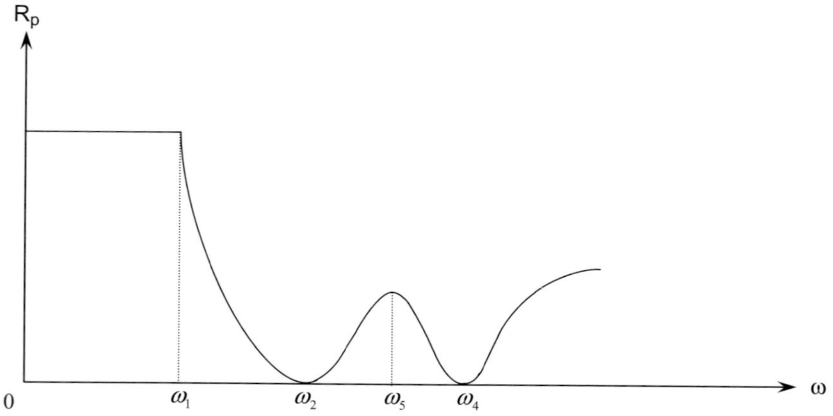

Fig. 4.3 Frequency dependence of the reflection coefficient R_p given by Eq. (4.22).

the function $R_p(\omega)$ has a maximum, which is equal to

$$R_{max} = \left(\frac{1 - \sin 2\psi}{1 + \sin 2\psi}\right)^2 \tag{4.25}$$

The Brewster angle is given by

$$\tan^2 \psi_B = \varepsilon_\perp(\omega) \tag{4.26}$$

and exists only if $\omega > \omega_{0\perp}$. Now let us note that, using the law of refraction $\sin \psi_1 = n^{-1} \sin \psi$, we can write Eq. (4.22) in the form

$$R_p = \frac{\tan^2(\psi - \psi_1)}{\tan^2(\psi + \psi_1)} \tag{4.22a}$$

At $\psi = \psi_B$, the angle of refraction is $\psi_1 = \frac{\pi}{2} - \psi_B$, and the reflected wave has no component of the electric field vector in the plane of incidence.

4.3 Influence of Phonon–Plasmon Coupling on Reflection From a Polar Uniaxial Semiconductor

In this section we shall discuss briefly the reflection of the waves which are incident normally on the surface of a uniaxial polar semiconductor, whose optical axis makes an angle θ with the normal (see Fig. 4.4). In this case the angle of reflection $\psi = 0$ and the coefficient of reflection is given by

$$R(\omega) = \left|\frac{1 - n}{1 + n}\right|^2 \tag{4.27}$$

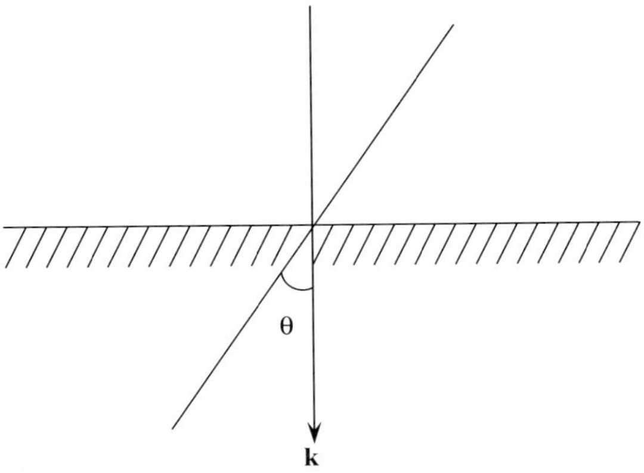

Fig. 4.4 Orientation of the crystal axis with respect to the reflecting surface and to the wave vector of the incident wave.

where $n = n_0$ given by Eq. (2.23), if the electric field vector $\boldsymbol{E}$ in the incident wave is perpendicular to the plane containing the wave vector and the crystal axis (case A), and $n = n_e$ given by Eq. (2.24), if $\boldsymbol{E}$ is parallel to that plane (case B).

In case A, the frequency dependence of the coefficient of reflection is analogous to that shown in Fig. 4.2. The regions of total reflection ($R = 1$) are

$$\omega < \omega_{\perp -}, \quad \omega_{\mathrm{T}\perp} < \omega < \omega_{\perp +} \tag{4.28}$$

In both transmission regions $R(\omega)$ decreases rapidly from 1 to zero. The frequencies at which $R(\omega) = 0$ are given by

$$\omega^2 = \frac{1}{2a_\perp^\infty}\left\{\omega_{0\perp}^2 + a_\perp^0\omega_{\mathrm{L}\perp}^2 \pm \left[\left(\omega_{0\perp}^2 + a_\perp^0\omega_{\mathrm{L}\perp}^2\right)^2 - 4a_\perp^\infty\omega_{0\perp}^2\omega_{\mathrm{T}\perp}^2\right]^{1/2}\right\} \tag{4.29}$$

where

$$a_\perp^{0,\infty} = 1 - \frac{1}{\varepsilon_\perp^{0,\infty}} \tag{4.29a}$$

These results are also valid in the case B, if $\theta = 0$ For $\theta = \pi/2$ the symbol $\perp$ in Eqs. (4.28) and (4.29) should be replaced by the symbol $\|$.

Let us now consider the function $R(\omega)$ in the case B at $\theta \neq 0, \pi/2$. The reflection of the waves in this case is considerably different as compared with that from the surface of an isotropic (cubic) crystal. Really, substituting Eq. (2.24) into Eq. (4.27) we find four total reflection regions and four regions of transmission. The doubling of the regions of transparency and nontransparency as compared with the reflection from the surface of a uniaxial nonpolar crystal is caused by the appearance of four branches of coupled phonon–plasmon polaritons. Low-frequency limits of the transmission regions coincide with the cut-off frequencies $\omega_{\mathrm{L}\pm}$ and $\Omega_{\mathrm{L}\pm}$ of the extraordinary wave described by Eq. (2.24), and high-frequency limits of those (for three low-frequency regions) coincide with the resonance frequencies, which are given by solutions of Eq. (2.20) and which are dependent essentially on the angle θ (see Fig. 2.3). The high-frequency transmission region is bounded only from below. The frequencies at which $R(\omega) = 0$ (and $n = 1$) for most semiconductors coincide practically with the cut-off frequencies of the phonon–plasmon polaritons.

4.4
Magnetoplasmon Reflection for the Faraday Configuration

In Chapter 3, we have neglected the reflection of the waves at the crystal surface. In this section, we shall study the reflection of a plane-polarized wave which is incident normally on a crystal's surface, but has an arbitrary orientation with respect to the crystal axis.

Using standard boundary conditions and Eq. (3.34) for $\chi_\pm$, we obtain the following expression for the reflection coefficient:

$$R = a_+ \left|\frac{1-n_+}{1+n_+}\right|^2 + a_- \left|\frac{1-n_-}{1+n_-}\right|^2 \tag{4.30}$$

where

$$a_\pm = \frac{1}{2}\left(1 \mp \frac{\eta \cos 2a}{\sqrt{1+\eta^2}}\right) \tag{4.31}$$

Here η is given by Eq. (3.36), a is the angle between the electric field vector in the incident wave and the plane containing the crystal axis and the wave vector, and $n_\pm$ are the refractive indices of the elliptically polarized waves which are defined by Eq. (3.19). For $a = \frac{\pi}{2}\left(l + \frac{1}{2}\right)$, Eq. (4.30) can be simplified:

$$R = \frac{|n_+ - n_-|^2 + |1 - n_+ n_-|^2}{|1+n_+|^2 |1+n_-|^2} \tag{4.32}$$

If also $\varphi = 0$ (the crystal axis is parallel to the constant magnetic field) and $\eta = 0$ then $n_\pm$ are given by Eq. (3.22). In this case an analysis of Eq. (4.32) indicates that, for $\omega_{0\perp}/\Omega < \sqrt{2}$, the medium is transparent in the whole frequency range; in the region

$$\Omega < \omega < \frac{\Omega}{2}\left(1 + \sqrt{1 + 4\frac{\omega_{0\perp}^2}{\Omega^2}}\right) \tag{4.33}$$

the medium is transparent only for the (+) wave, whereas for

$$\omega < \frac{\Omega}{2}\left(\sqrt{1 + 4\frac{\omega_{0\perp}^2}{\Omega^2}} - 1\right) \tag{4.34}$$

it is transparent only for the (–) wave. For $\omega_{0\perp}/\Omega > \sqrt{2}$, the transparency and nontransparency regions alternate when ω is varied. Total reflection occurs in the range

$$\Omega < \omega < \frac{\Omega}{2}\left(\sqrt{1 + 4\frac{\omega_{0\perp}^2}{\Omega^2}} - 1\right) \tag{4.35}$$

but at other frequencies the crystal is transparent. In the case when only one of the waves can propagate in the crystal, Eq. (4.32) is replaced by

$$R = \frac{1+n^2}{(1+n)^2} \tag{4.36}$$

which has a minimum $R_{\min}= 0.5$ at $n= 1$. For $\varphi = 0$ the minimum occurs at

$$\omega = \frac{\Omega}{2}\left(\sqrt{1+\frac{4\omega_{0\perp}^2}{\Omega^2\left[1-(\varepsilon_\perp^\infty)^{-1}\right]}}+1\right) \tag{4.37a}$$

if the (–) wave is propagated, and at

$$\omega = \frac{\Omega}{2}\left(\sqrt{1+\frac{4\omega_{0\perp}^2}{\Omega^2\left[1-(\varepsilon_\perp^\infty)^{-1}\right]}}-1\right) \tag{4.37b}$$

if the (+) wave is propagated. Note that the distance between these two values of frequency coincides with the cyclotron frequency $\Omega = \frac{eB_0\mu_\perp}{mc}$. When $\omega << \Omega << \omega_{0\perp}$, which corresponds to the propagation of helicons [43, 44], the medium is transparent only for the (–) wave with a very large refractive index. In this case, we obtain

$$R \approx 1 - \frac{2\sqrt{\omega\Omega}}{\omega_{0\perp}\sqrt{\varepsilon_\perp^\infty}} \tag{4.38}$$

Let us now consider the polarization of the reflected wave. In the coordinate system x', y', z, in which the axis z is parallel to $\mathbf{B}_0$ and x' is parallel to the polarization plane of the incident wave, the following relation for the reflected wave is easily obtained:

$$\frac{E_{y'}}{E_{x'}} = \frac{\mathrm{i}+\eta\sin 2a}{r\sqrt{1+\eta^2}-\eta\cos 2a} \tag{4.39}$$

where

$$r \equiv \frac{n_+n_- - 1}{n_+ - n_-} \tag{4.40}$$

Equation (4.32) implies that the reflected wave is elliptically polarized and that the major axis of the polarization ellipse does not coincide with the direction of the electric field vector in the incident wave. In the frequency region in which both waves can propagate ($n_\pm^2 > 0$), the rotation angle ψ and the ellipticity ξ are given by

$$\tan 2\psi = \frac{2\eta\left[r\sqrt{1+\eta^2} - \eta\cos 2a\right]\sin 2a}{\left[r\sqrt{1+\eta^2} - \eta\cos 2a\right]^2 - 1 - \eta^2\sin^2 2a} \tag{4.41}$$

$$\frac{\xi}{1+\xi^2} = \frac{r\sqrt{1+\eta^2} - \eta\cos 2a}{(1+r^2)(1+\eta^2) - 2r\eta\sqrt{1+\eta^2}\cos 2a} \tag{4.42}$$

Note that for $\varphi = 0$ we have $\psi = 0$ and $\xi = r$ as in the case of an isotropic medium. For $\varphi \neq 0$ we have $\psi = 0$ only for $a = l\pi/2,\ l = 0, \pm 1, \ldots$ that is, if the polarization plane of the incident wave is parallel or perpendicular to the plane containing $\boldsymbol{B}_0$ and the crystal axis.

Thus, for $\varphi \neq 0$ and $a \neq l\pi/2$ the rotation of the principal axes of the polarization ellipse in the reflected wave arises with respect to the polarization plane of the incident wave. It is important that this effect is only possible if the reflection takes place from the surface of a uniaxial semiconductor.

4.5
Magnetoplasmon Reflection for the Voigt Configuration

Let us assume now that a static magnetic field $\boldsymbol{B}_0$ is parallel to the reflecting surface and the electromagnetic wave is incident normally from the vacuum. We shall restrict ourselves to the following particular cases of the crystal-axis (**C**) orientation:

A. **C** is perpendicular to the reflecting surface.
B. **C** is parallel to the reflecting surface and perpendicular to $\boldsymbol{B}_0$.
C. **C** is parallel to $\boldsymbol{B}_0$.

In the case A, the reflection coefficient is given by

$$R = \left|\frac{1 - \sqrt{\varepsilon_\perp^\infty\left(1 - \omega_{0\perp}^2/\omega^2\right)}}{1 + \sqrt{\varepsilon_\perp^\infty\left(1 - \omega_{0\perp}^2/\omega^2\right)}}\right|^2 \tag{4.43}$$

if the electric field vector $\boldsymbol{E}$ in the incident wave is parallel to $\boldsymbol{B}_0$. Thus, in this case R is independent of B_0, and the transmission occurs only in the region $\omega > \omega_{0\perp}$. At the frequency

$$\omega = \frac{\omega_{0\perp}}{\sqrt{1 - \left(\varepsilon_\perp^\infty\right)^{-1}}} \tag{4.44}$$

we find $R = 0$. For $\boldsymbol{E} \perp \boldsymbol{B}_0$, we obtain

$$R = \left|\frac{1-n}{1+n}\right|^2 \tag{4.45}$$

where

$$n^2 = \varepsilon_\perp^\infty \frac{(\omega^2 - \omega_+^2)(\omega^2 - \omega_-^2)}{\omega^2\left(\omega^2 - \Omega^2 - \omega_{0\|}^2\right)} \tag{4.46}$$

is the refractive index of the extraordinary wave and $\omega_\pm$ are given by Eq. (3.74). Total reflection occurs in the frequency regions

$$\omega < \omega_-, \qquad \sqrt{\Omega^2 + \omega_{0\|}^2} < \omega < \omega_+; \tag{4.47}$$

At any other frequencies transmission takes place. At frequencies

$$\omega_{1,2} = \frac{1}{\sqrt{2}}\left\{\Omega^2 + \omega_{0\|}^2 + \frac{\omega_{0\perp}^2}{1-(\varepsilon_\perp^\infty)^{-1}} \pm \left[\left(\Omega^2 + \omega_{0\|}^2 + \frac{\omega_{0\perp}^2}{1-(\varepsilon_\perp^\infty)^{-1}}\right)^2 - 4\frac{\omega_{0\|}^2\omega_{0\perp}^2}{1-(\varepsilon_\perp^\infty)^{-1}}\right]^{1/2}\right\}^{1/2} \tag{4.48}$$

the reflection coefficient has minima. Both ω_1 and ω_2 are changed essentially when the magnetic field is varied. In Fig. 4.5, $\omega_{1,2}$ are plotted as functions of Ω; the distance between ω_1 and ω_2 is increasing monotonically with increasing magnetic field.

In the case B, the reflection coefficient coincides with Eq. (4.43) if the electric field vector in the incident wave is parallel to $\boldsymbol{B}_0$. If $\boldsymbol{E}$ is parallel to the crystal axis, R is given by Eq. (4.45), where

$$n^2 = \varepsilon_\|^\infty \frac{(\omega^2 - \omega_+^2)(\omega^2 - \omega_-^2)}{\omega^2(\omega^2 - \Omega^2 - \omega_{0\perp}^2)} \tag{4.49}$$

In this case the medium is transparent in the following two frequency regions:

$$\omega_- < \omega < \sqrt{\Omega^2 + \omega_{0\perp}^2}, \quad \omega < \omega_+ \tag{4.50}$$

With increasing ω in both regions, $R(\omega)$ decreases from 1 until zero and then increases. The frequencies at which $R(\omega) = 0$ differ from those given by Eq. (4.48) only by replacing the signs $\perp$ and $\|$.

In the case C, the reflection coefficient is independent of the magnetic field, if $\boldsymbol{E}\|\boldsymbol{B}_0$, and coincides with Eq. (4.43), where the sign $\perp$ should be replaced by the sign $\|$. Transmission takes place only at $\omega > \omega_{0\|}$; $R(\omega)$ vanishes at

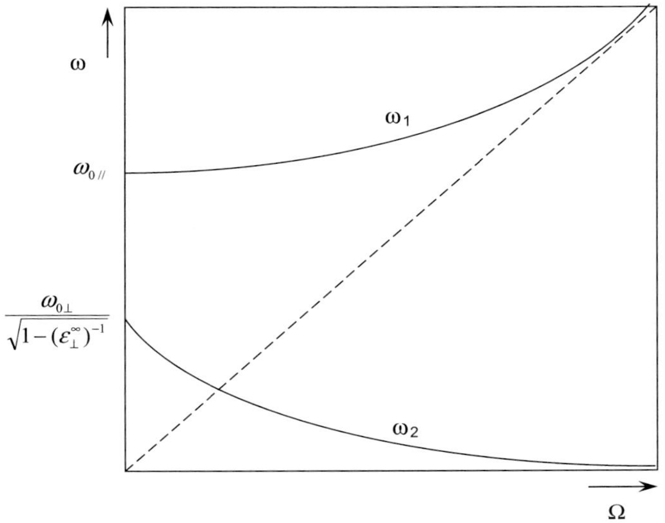

Fig. 4.5 Magnetic field dependence of the frequencies ω_1 and ω_2 given by Eq. (4.48). It is assumed that $\frac{\mu_{||}}{\varepsilon_{||}^{\infty}} > \frac{\mu_{\perp}}{\varepsilon_{\perp}^{\infty} - 1}$.

$$\omega = \frac{\omega_{0||}}{\sqrt{1 - \left(\varepsilon_{||}^{\infty}\right)^{-1}}} \tag{4.51}$$

If $\boldsymbol{E} \perp \boldsymbol{B}_0$, the reflection coefficient can be expressed by Eq. (4.45), where n is the refractive index of the extraordinary wave defined by Eq. (3.66). Total reflection occurs in two regions:

$$\omega < \omega_-, \quad \sqrt{\Omega^2 + \omega_{0\perp}^2} < \omega < \omega_+. \tag{4.52}$$

where $\omega_\pm$ now are the cut-off frequencies given by Eq. (3.67). In each of the two transmission regions, $R(\omega)$ decreases with increasing ω, vanishes and then increases. The frequencies at which $R(\omega) = 0$ are given by

$$\omega = \frac{1}{2}\left\{\Omega^2 + \omega_{0\perp}^2\left(1 + \frac{\varepsilon_\perp^\infty}{\varepsilon_\perp^\infty - 1}\right) \pm \left[(\Omega^2 + \omega_{0\perp}^2)^2 + \omega_{0\perp}^4\left(\frac{\varepsilon_\perp^\infty - 2}{\varepsilon_\perp^\infty - 1}\right)^2 + 2(\Omega^2 + \omega_{0\perp}^2)\frac{\varepsilon_\perp^\infty\,\omega_{0\perp}^2}{(\varepsilon_\perp^\infty - 1)}\right]^{1/2}\right\} \tag{4.53}$$

Note that in all the above-considered cases A, B and C the reflected wave is polarized in the same plane in which the incident wave is polarized.

Part 2
Surface and Interface Electromagnetic Waves in Semiconductor Structures

Roland H. Tarkhanyan

Radiowaves and Polaritons in Anisotropic Media. Roland H. Tarkhanyan and Nikolaos K. Uzunoglu

ISBN: 3-527-40615-8

Introduction

One of the most actual directions of modern solid-state optics is associated with the spectroscopy of surface polaritons [35–40]. Surface electromagnetic waves propagate along a crystal surface and are described by solutions of the Maxwell equations, for which the fields are exponentially decaying for increasing distances from the boundary between the crystal and a vacuum. Interface waves are a simple generalization of the surface waves, when the second medium is not a vacuum and the waves propagate along the boundary between two media. Surface polaritons for isotropic (cubic) polar crystals have been studied in Refs. [41–45] and for anisotropic media in Refs. [46–51]. The main experimental method for observation of the surface waves is the attenuated total reflection (ATR) method [35]. For experimental aspects of detection of the surface waves we refer the reader to [40, 52–56] and the literature cited therein.

In Part 2 of the present book, we shall investigate phonon–plasmon polaritons bound to the surface of a uniaxial polar semiconductor. We shall see that in comparison to bulk polaritons, the surface polaritons have very different properties, which depend on boundary conditions and require special theoretical consideration [50, 51].

We shall assume that the spatial dispersion as well as the variation of the dielectric permittivity near the crystal surface due to the microscopic effects may be neglected. These assumptions are reasonable for polaritons whose wavelength is large compared to the lattice spacing. We shall also neglect the damping processes, assuming that $\omega\tau >> 1$.

All basic characteristics of surface phonon–plasmon polaritons in a semi-infinite uniaxial polar semiconductor and in a slab of finite thickness are studied in Chapters 5 and 6, respectively. They include some new phenomena caused by anisotropy; for example, spatial oscillations of the wave amplitude, the dependence of the surface branches' number on the free carrier density, and so forth.

Interface magnon–plasmon polaritons between a uniaxial antiferromagnetic insulator and a semiconductor are considered in Chapter 7. It is well known that only TM-type interface polaritons can travel along a flat interface between two nonmagnetic media [40]. On the other hand, only TE-type polaritons exist on the interface between a magnetic and a nonmagnetic material in the absence of the free charge carriers [57]. We shall see that nonradiative interface polaritons of

Radiowaves and Polaritons in Anisotropic Media. Roland H. Tarkhanyan and Nikolaos K. Uzunoglu

ISBN: 3-527-40615-8

both TE and TM types can travel along a semiconductor/antiferromagnetic insulator interface, which cannot occur in the other aforementioned structures [58]. This is a particularly important factor from the experimental point of view, since it opens a possibility of studying different types of the waves in the same structure. Secondly, we shall show that resonant excitation of the interface polaritons in a semiconductor/antiferromagnetic insulator two-layer structure leads to a very interesting new phenomenon, namely, the complete transmission of obliquely incident electromagnetic waves by the structure, for waves whose frequency lies outside the transmission range for one of the layers [58]. It is the excitation of the interface polaritons that mediates the transfer of the incident wave's energy across a nontransparent region.

It should be noted that a similar effect has so far been considered only in plasma [59, 60]. It has been shown that an inhomogeneous layer of dense plasma can become completely transparent only for a p-polarized incident wave. In Ref. [61] it has been reported that a two-layer structure consisting of an antiferromagnetic and a nonmagnetic insulator can be completely transparent for an s-polarized incident wave, whereas it does not transmit p-polarized waves [61]. In contrast to the cases studied in Refs. [59–61], complete transmission by a two-layered structure consisting of a semiconductor and an antiferromagnetic insulator is possible for both s- and p-polarized incident waves, and it is accompanied by many other interesting effects, too. These effects can be used not only for the excitation of interface polaritons but also to produce polarizers and filters for electromagnetic waves in the infrared range.

Finally, in Chapter 8 we will consider the coupled surface polaritons which propagate along the lateral surfaces of a ferrite/semiconductor superlattice, in the presence of a nonquantizing static magnetic field [77], as well as in the quantum Hall effect conditions, when a quantizing magnetic field is perpendicular to the quantum well planes [78]. Collective excitations in superlattices that consist of alternating layers of two or three different materials have been studied in a many works (for example, in Refs. [79, 80], etc.). We shall restrict ourselves to the consideration of coupled magnon–plasmon surface polaritons, that is, of the nonradiative spin–electromagnetic waves.

5
Surface Polaritons in Uniaxial Semiconductors

5.1
General Dispersion Relation of Polaritons Bound to the Surface of a Semi-Infinite Semiconductor

In this chapter we will consider surface electromagnetic waves for the example of a uniaxial semiconductor under conditions when dissipation processes, spatial dispersion and also transitions associated with a change in the number of free carriers can be ignored. We shall show that these localized waves have a number of interesting and unexpected features that are not present in the case of isotropic (cubic) crystals.

Suppose that the semiconductor occupies the space region $z<0$; the region $z>0$ is taken to be a vacuum. The crystal axis lies in the xz-plane, making an angle φ with the z-axis. In this coordinate system, the permittivity tensor has the form

$$\varepsilon = \begin{pmatrix} \varepsilon_{xx} & 0 & \varepsilon_{xz} \\ 0 & \varepsilon_{yy} & 0 \\ \varepsilon_{zx} & 0 & \varepsilon_{zz} \end{pmatrix}, \tag{5.1}$$

where

$$\varepsilon_{xx} = \varepsilon_{\perp} \cos^2 \varphi + \varepsilon_{\|} \sin^2 \varphi, \quad \varepsilon_{zz} = \varepsilon_{\perp} \sin^2 \varphi + \varepsilon_{\|} \cos^2 \varphi \tag{5.1a}$$

$$\varepsilon_{yy} = \varepsilon_{\perp}, \qquad \varepsilon_{xz} = \varepsilon_{zx} = (\varepsilon_{\|} - \varepsilon_{\perp}) \sin \varphi \cos \varphi \tag{5.1b}$$

Here $\varepsilon_{\perp}$ and $\varepsilon_{\|}$ are the principal values of the permittivity tensor defined by Eqs. (1.30) and (1.31) for a nonpolar semiconductor and by Eqs. (2.18a) and (2.18b) for a polar one.

Theory of the surface electromagnetic waves reduces to finding solutions of the Maxwell equations in the form of inhomogeneous plane waves which decrease with increasing distance from the surface:

Radiowaves and Polaritons in Anisotropic Media. Roland H. Tarkhanyan and Nikolaos K. Uzunoglu

ISBN: 3-527-40615-8

$\boldsymbol{E}, \boldsymbol{H} \sim \exp[\mathrm{i}(\boldsymbol{kr} - \omega t)]$, where the tangential components of the wave vector k_x and k_y are real while the normal component k_z is a complex quantity. Therefore, for the region $z > 0$ (vacuum) we have to choose a solution for which Im $k_z > 0$ and for the region $z < 0$ (crystal) a solution with Im $k_z < 0$. Thus, the field of the surface wave in the vacuum can be written as

$$\boldsymbol{E}(\boldsymbol{r}, \boldsymbol{t}) = \boldsymbol{E}\mathrm{e}^{\mathrm{i}(k_x x + k_y y - \omega t) - \alpha z}, \quad \boldsymbol{H} = (\mu_0 \omega)^{-1}[\boldsymbol{kE}] \tag{5.2}$$

where

$$k_z \equiv \mathrm{i}\,\alpha, \qquad \alpha = \sqrt{k_x^2 + k_y^2 - \frac{\omega^2}{c^2}} \tag{5.3}$$

In the region of the semiconductor the surface wave is, in general, a linear combination of ordinary (o) and extraordinary (e) waves, whose fields have the form (the exponential term is omitted)

$$\boldsymbol{E}_\mathrm{o} = E[\boldsymbol{k}_\mathrm{o}\boldsymbol{a}] \quad \boldsymbol{H}_0 = (\mu_0 \omega)^{-1} [\boldsymbol{k}_\mathrm{o} \boldsymbol{E}_\mathrm{o}] \tag{5.4}$$

$$\boldsymbol{H}_\mathrm{e} = H[\boldsymbol{a}\boldsymbol{k}_\mathrm{e}] \quad \boldsymbol{E}_\mathrm{e} = -(\varepsilon_0 \omega)^{-1} \varepsilon^{-1} [\boldsymbol{k}_\mathrm{e} \boldsymbol{H}_\mathrm{e}] \tag{5.5}$$

where $\boldsymbol{a} = \{\sin\varphi, 0, \cos\varphi\}$ is a unit vector along the crystal axis.

Using the general dispersion relation (1.37), we obtain

$$n_\mathrm{o}^2 = \varepsilon_\perp, \qquad n_\mathrm{e}^2 = \frac{\varepsilon_\perp \varepsilon_{||}}{\varepsilon_{xx} s_x^2 + \varepsilon_{zz} s_z^2 + \varepsilon_{yy} s_y^2 + 2\varepsilon_{xz} s_x s_z} \tag{5.6}$$

from which the following expressions for the normal wave-vector components k_{oz} and k_{ez} can be found:

$$k_{oz} = -\mathrm{i}\,\alpha_\mathrm{o}, \qquad k_{ez} = \alpha_1 - \mathrm{i}\,\alpha_\mathrm{e} \tag{5.7}$$

where

$$\alpha_\mathrm{o} \equiv \sqrt{k_{||}^2 - \varepsilon_\perp \frac{\omega^2}{c^2}}, \qquad \alpha_1 = -k_x \frac{\varepsilon_{xz}}{\varepsilon_{zz}} \tag{5.8}$$

$$\alpha_\mathrm{e} = \frac{\sqrt{\varepsilon_\perp [\varepsilon_{||} k_x^2 + \varepsilon_{zz} \left(k_y^2 - \varepsilon_{||}\omega^2/c^2\right)}}{|\varepsilon_{zz}|} \tag{5.9}$$

and

$$k_{||}^2 \equiv k_x^2 + k_y^2 \tag{5.10}$$

Nonradiative surface waves exist only when a, a_0 and a_e are real and positive. The dispersion relation of the surface waves, which establishes a connection between the frequency and the two-dimensional wave vector $\boldsymbol{k}_{||} = \{k_x, k_y\}$, is obtained by using the standard boundary conditions for the tangential components of the fields at the plane $z = 0$:

$$H_x = H_{ox} + H_{ex}, \qquad E_x = E_{ox} + E_{ex},$$
$$H_y = H_{oy} + H_{ey}, \qquad E_y = E_{oy} + E_{ey}, \tag{5.11}$$

Eliminating from Eqs. (5.11) the field components in vacuum, we can rewrite the boundary conditions as two equations for the fields of the ordinary and extraordinary waves at $z = 0$:

$$k_x k_y (E_{ox} + E_{ex}) + \left(k_y^2 - a^2\right)\left(E_{oy} + E_{ey}\right) + \mathrm{i}\, a\mu_0\omega(H_{ox} + H_{ex}) = 0$$
$$(k_x^2 - a^2)(E_{ox} + E_{ex}) + k_x k_y \left(E_{oy} + E_{ey}\right) - \mathrm{i}\, a\mu_0\omega\left(H_{oy} + H_{ey}\right) = 0 \tag{5.12}$$

Using Eqs. (5.4) and (5.5), we obtain from Eqs. (5.12) a system of two homogeneous equations for E and H, whose solvability condition gives the dispersion relation

$$\left(k_x k_{oz} \sin\varphi - k_{||}^2 \cos\varphi\right)\left[(\varepsilon_\perp - 1)k_x \sin\varphi - (k_{ez} - \mathrm{i}\, a\varepsilon_\perp)\cos\varphi\right] - \varepsilon_\perp(\mathrm{i}\, a - k_{ez})\sin\varphi$$
$$\left[\mathrm{i}\, a k_x \cos\varphi + \left(k_y^2 - \mathrm{i}\, a k_{oz}\right)\sin\varphi\right] = 0 \tag{5.13}$$

Substitution of Eqs. (5.7) into Eq. (5.13) reduces the latter to the form $A + \mathrm{i}B = 0$, where A and B are real functions of ω, φ, k_x and k_y. For real values of k_x and k_y the frequency ω of the surface wave is, in general, complex. In what follows, we shall only investigate the stationary nonradiative surface waves, for which ω is real. Analysis of Eq. (5.13) shows that for an arbitrary direction of propagation, such surface waves exist only in the cases $\varphi = 0$ (e-wave) and $\varphi = \pi/2$ (combination of both modes). For $\varphi \neq 0, \pi/2$ they exist only in the case of propagation perpendicular ($k_x = 0$) or parallel ($k_y = 0$) to the plane passing through the crystal axis and the normal to the surface. In the first case, the wave is a combination of the ordinary and extraordinary modes; in the second case, the wave is an extraordinary one.

5.2
Amplitude Oscillations of the Surface Waves

It can easily be seen that when $k_y = 0$ the field of the extraordinary wave decreases with the distance into the crystal, executing spatial oscillations. The characteristic oscillation length

$$l = \frac{2\pi}{|\operatorname{Re} k_{ez}|} = \frac{2\pi}{k_{\|}} \left| \frac{\varepsilon_{zz}}{\varepsilon_{xx}} \right| \tag{5.14}$$

depends on the frequency ω, the wave number $k_{\|}$ and the angle φ, which are related by the dispersion equation

$$\frac{c^2 k_{\|}^2}{\omega^2} = \frac{\varepsilon_{zz} - \varepsilon_{\|}\varepsilon_{\perp}}{1 - \varepsilon_{\|}\varepsilon_{\perp}}. \tag{5.15}$$

The oscillation regime of the amplitude of the surface waves can be detected if these waves are excited on the surface of a slab whose thickness d is comparable with the penetration depth $\lambda = |\alpha_e|^{-1}$, if $\lambda/l > 1$. In this case, the wave has a finite amplitude for $|z| \leq d$ and in addition there is a reflected wave. Standard boundary conditions lead to the following expression for the tangential component of the electric field of the resultant wave within the crystal, for $z \geq -d$.

$$E_x = E_{ex} e^{i(k_{\|}x - \omega t)} \left[e^{ik_{ez}z} - r e^{-ik_{ez}(z+2d)} \right] \tag{5.16}$$

where

$$r = \frac{a-b}{a+b}, \quad a = k_{\|}(\varepsilon_{\perp}k_z - k_{ez}) + \left(\varepsilon_{\perp}\omega^2/c^2 - k_{\|}^2\right)\tan\varphi, \quad b = \varepsilon_{\perp}k_z k_{ez}\tan\varphi. \tag{5.17}$$

On the front surface $z = 0$, the reflected wave is exponentially small compared with the primary wave, so that we shall ignore it; but on the rear surface $z = -d$ the field of the wave must oscillate when the angle φ, ω or $k_{\|}$ is varied.

For $\varphi = 0$ and $\varphi = \pi/2$ as well as in the isotropic case, there are no oscillations. In the next section, we shall analyze these special cases.

5.3
Peculiarities of Surface Polaritons in Uniaxial Polar Semiconductors in Some Special Cases

In the case $\varphi = 0$ when the crystal axis is perpendicular to the surface, Eq. (5.13) gives

$$\alpha_e = -\varepsilon_{\perp}\alpha \tag{5.18}$$

Using Eqs. (5.3), (5.9) and (5.18), we find that the surface electromagnetic wave is characterized by the dispersion relation

$$\frac{k_{\|}^2 c^2}{\omega^2} = \frac{\varepsilon_{\|}(1 - \varepsilon_{\perp})}{1 - \varepsilon_{\|}\varepsilon_{\perp}} \tag{5.19}$$

and by the inequalities

$$\varepsilon_\perp < 0, \quad k_{||} > \frac{\omega}{c}, \quad \varepsilon_{||}\left(\frac{k_{||}^2 c^2}{\omega^2} - \varepsilon_{||}\right) < 0. \tag{5.20}$$

The last condition in conjunction with Eq. (5.19) is satisfied either for $\varepsilon_{||} > 1$ or for $\varepsilon_{||} < -|\varepsilon_\perp|^{-1}$.

In the case $\varphi = \pi/2,\ k_y = 0$, the dispersion relation differs from Eq. (5.19) only in the replacement of $\varepsilon_\perp$ by $\varepsilon_{||}$, and *vice versa*.

In the case $\varphi = \pi/2,\ k_x = 0$ the surface wave is an ordinary one; it is characterized by

$$\frac{k_{||}^2 c^2}{\omega^2} = \frac{\varepsilon_\perp}{\varepsilon_\perp + 1}, \quad \varepsilon_\perp < -1 \tag{5.21}$$

In this special case, the anisotropy is unimportant and, as in the case of an isotropic polar semiconductor, there are two branches of the surface phonon–plasmon polaritons, which are shown in Fig. 5.1. The high-frequency branch begins at the point $\omega = k_{||}c = \omega_{T\perp}$; its frequency increases with increasing $k_{||}$, and for $k_{||} \gg \omega/c$ approaches asymptotically the value

$$\omega^2 = \frac{\varepsilon_{\perp\infty}}{2(1+\varepsilon_{\perp\infty})}\left\{\omega_{L\perp}^2 + \omega_{0\perp}^2 + \frac{\omega_{T\perp}^2}{\varepsilon_{\perp\infty}} + \left[\left(\omega_{L\perp}^2 + \omega_{0\perp}^2 + \frac{\omega_{T\perp}^2}{\varepsilon_{\perp\infty}}\right)^2 - 4\omega_{0\perp}^2\omega_{T\perp}^2\left(1 + \frac{1}{\varepsilon_{\perp\infty}}\right)\right]^{1/2}\right\}, \tag{5.22}$$

which corresponds to the solution of the dispersion relation in the absence of retardation, $\varepsilon_\perp = -1$. The low-frequency branch exists for all values of $k_{||}$; the

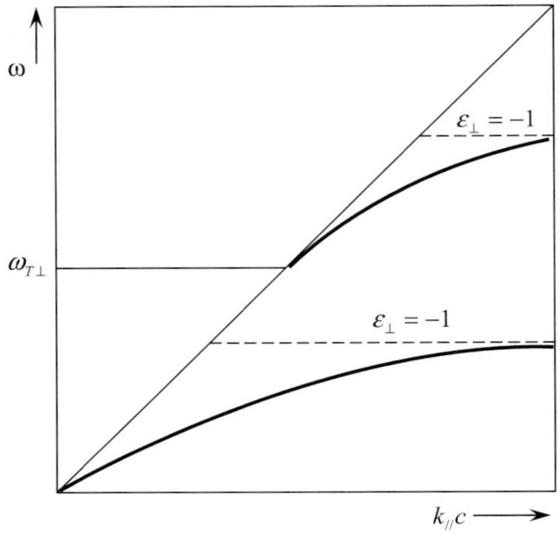

Fig. 5.1 Dispersion curves of the surface polaritons in the case $\varphi = \pi/2$, $k_x = 0$ (ordinary wave).

frequency increases with increasing $k_{||}$ from zero to a value which differs from that given by Eq. (5.22) only by the sign in front of the radical in the square brackets.

In the case $\varphi = 0$, the situation differs qualitatively from the isotropic case since there are new surface polariton branches whose number and regions of existence depend on the density of the free charge carriers and on the relative positions of the characteristic frequencies $\omega_{T\perp}, \omega_{T||}, \omega_{L\perp}, \omega_{L||}, \omega_0, \omega_\perp^{+,-}, \omega_{||}^{+,-}, \Omega_\perp^{+,-}, \Omega_{||}^{+,-}$, where the last eight values are determined by the relations

$$\varepsilon_\perp\left(\omega_\perp^{+,-}\right) = 0, \qquad \varepsilon_{||}\left(\omega_{||}^{+,-}\right) = 0, \qquad \varepsilon_\perp\left(\Omega_\perp^{+,-}\right) = 1, \qquad \varepsilon_{||}\left(\Omega_{||}^{+,-}\right) = 1 \quad (5.23)$$

For a given value of $k_{||}$, the maximum number of branches is four. The curves $\omega = \omega_i(k_{||})$ corresponding to the solutions of Eq. (5.19) begin at the points $\omega_{||}^{-1}, \omega_{||}^{+}$, $\Omega_\perp^-$ and $\Omega_\perp^+$ for $k_{||} = 0$, increase with increasing $k_{||}$ and for $k_{||} >> \omega/c$ approach asymptotically the frequencies of the surface modes in the absence of retardation, which are given by solutions of the equation

$$\varepsilon_\perp \varepsilon_{||} = 1 \quad (5.24)$$

However, nonradiative surface waves correspond only to those sections of these curves that lie in the regions of the $(\omega, k_{||})$-plane in which the conditions (5.20) are satisfied. Equation (5.24) is of the fourth order in ω^2, but only one, two or three solutions lie in the regions in which $\varepsilon_\perp < 0$. Thus, three, two or one of the dispersion curves terminate at finite values of $k_{||}$, which is impossible in the isotropic case. The number of these solutions depends on the free carrier density. For example, in the case of a crystal with characteristic frequencies $\omega_{T\perp} < \omega_{L\perp} < \omega_{T||} < \omega_{L||}$ and for

$$\omega_\perp^+ < \omega_{T||}, \quad \omega_{||}^- < \omega_{T\perp} \quad (5.25)$$

there are three such solutions. We denote by m the smallest and by M the largest of the following quantities:

$$\omega_1^2 = \omega_{T||}^2 \frac{\varepsilon_{\perp\infty}\left(\omega_{T||}^2 - \omega_{L\perp}^2\right)}{\mu_\perp\left(\omega_{T||}^2 - \omega_{T\perp}^2\right)} \qquad \omega_2^2 = \omega_{T\perp}^2 \frac{\varepsilon_{||\infty}\left(\omega_{L||}^2 - \omega_{T\perp}^2\right)}{\mu_{||}\left(\omega_{T||}^2 - \omega_{T\perp}^2\right)} \quad (5.26)$$

Inequalities (5.25) are only satisfied when $\omega_0^2 \le m$. If the electron density is increased, when $m < \omega_0^2 < M$, one of the conditions (5.25) is violated and there then remain two branches, which terminate at finite values of $k_{||}$. For $\omega_0^2 > M$, both conditions (5.25) are violated and there then remains only one such branch. With increasing density of the conduction electrons, not only the number of surface modes but also the frequency of each of the modes changes, remaining always less than min $\{\omega_\perp^+, \omega_{||}^+\}$.

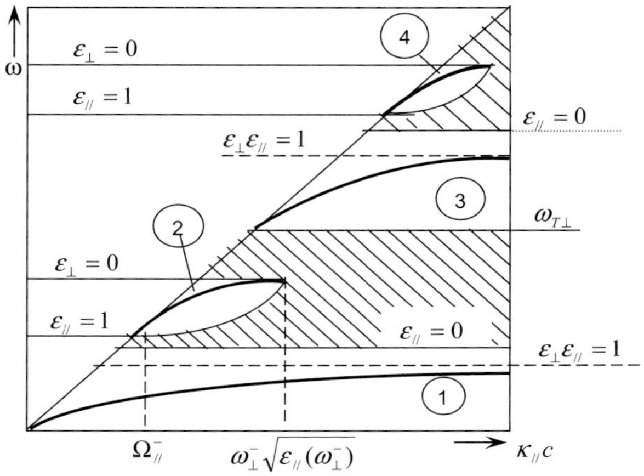

Fig. 5.2 Dispersion curves of the extraordinary surface waves in the case $\varphi = 0$, $\omega_{T||} < \omega_{T\perp} < \omega_{L||} < \omega_{L\perp}$, $\omega_0 < \omega_1$.

Figures 5.2 and 5.3 show the regions in the $(\omega, k_{||}c)$-plane in which the conditions (5.20) are satisfied, and the dispersion curves of the surface phonon–plasmon polaritons in the case $\varphi = 0$ for $\omega_{T||} < \omega_{T\perp} < \omega_{L||} < \omega_{L\perp}$ (such a case is realized, for example, in CdS). The dispersion curves begin at the points $\varepsilon_{\perp} = \infty$ $(\omega = 0, \omega = \omega_{T\perp})$ and $\varepsilon_{||} = 1$ $(\omega = \Omega_{||}^{+}, \omega = \Omega_{||}^{-})$ lying on the straight line $\omega = k_{||}c$. In the hatched regions, as in the region $\omega > k_{||}c$, nonradiative surface waves are impossible. Of particular interest are the branches lying in regions 2 and 4, since these branches are only possible in the anisotropic case. These regions are bounded below by the curve

$$\frac{k_{||}^2 c^2}{\omega^2} = \varepsilon_{||} \tag{5.27}$$

and to the left and above by the straight lines $\omega = k_{||}c$ and $\varepsilon_{\perp} = 0$, respectively. Figure 5.2 corresponds to the case $\omega_0 < \omega_1$ $(\omega_{\perp}^{-} < \omega_{T||})$. When this condition is violated as the density N increases, the dispersion curve in region 2 exists for all values $k_{||} > \Omega_{||}^{-}/c$, and the number of surface modes in the region

$$k_{||}c > \omega_{\perp}^{-}\sqrt{\varepsilon_{||}(\omega_{\perp}^{-})} \tag{5.28}$$

is increased by unity (Fig. 5.3). The dispersion curve in region 4 in both cases lies in the restricted range of $k_{||}$:

$$\Omega_{||}^{+} < k_{||}c < \omega_{\perp}^{+}\sqrt{\varepsilon_{||}(\omega_{\perp}^{+})} \tag{5.29}$$

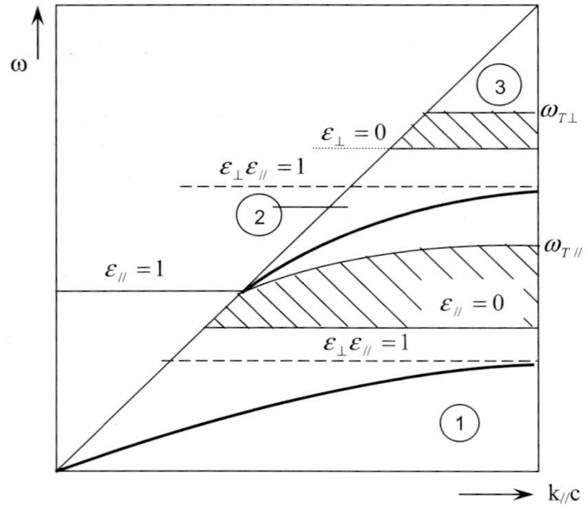

Fig. 5.3 Change of the dispersion curve in region 2 with increasing density of the conduction electrons ($\omega_O > \omega_1$).

The corresponding mode exists only if

$$\Omega_{||}^{+} < \omega < \omega_{\perp}^{+} \tag{5.30}$$

We conclude this chapter by enumeration of the main characteristic properties of the surface polaritons caused by anisotropy. Briefly, they are as follows.

1. In the isotropic case there are two branches of surface phonon–plasmon polaritons, one of which exists in the whole range of the two-dimensional wave vector along the surface ($k_{||}$)from 0 to ∞, and the second only existing for $k_{||} > \omega/c$. In contrast, in a uniaxial polar semiconductor four branches are possible. For large values $k_{||} >> \omega/c$ their number depends on the density (N) of the conduction electrons; it increases with increasing N and can take values from 1 to 3 for different positions of characteristic frequencies of the lattice optical vibrations. It also depends on the orientation of the distinguished axis of the crystal relative to its surface. In the opposite case $k_{||} < \omega/c$, as in the isotropic case, there remains only a single surface mode, namely, one of those that are also present when $k_{||} >> \omega/c$.
2. The remaining three modes arise at finite values of $k_{||} = \omega/c$ and exist either for an unbounded increase of $k_{||}$, as in the isotropic case, or in a bounded range of $k_{||}$ values; boundaries of the interval also depend on the conduction electron density N. On the boundaries of the region of existence, the penetration depth of the wave is either zero or infinite; the latter possibility means that the surface wave becomes a volume one.

3. In general, the surface waves are a superposition of the ordinary and extraordinary waves, in which the electric or magnetic field vector, respectively, is perpendicular to the plane containing the crystal axis and the wave vector. In the special case when the axis of the crystal is parallel to its surface and the wave propagates at right angles to the axis, the surface wave is an ordinary one and its properties (in particular, the number of branches) are the same as in the isotropic case. But, if the wave propagates along the axis, or the axis is perpendicular to the surface, the wave is an extraordinary one. Another case when the wave is extraordinary is when it propagates parallel to the plane containing the crystal axis and the normal to the surface. The features mentioned above are present only when the wave contains an extraordinary component.
4. If the crystal axis is neither perpendicular nor parallel to the vector normal to the surface, nonradiative surface waves exist only in the case of propagation perpendicular or parallel to the plane of these vectors; in the latter case, the field of the wave decreases inwards and is oscillating. Under certain conditions, there are a large number of oscillations within the penetration depth; the characteristic length of the oscillations depends on the orientation of the surface, the plasma frequency and the frequency of the wave.

6
Surface Waves in a Uniaxial Semiconductor Slab

6.1
General Theory

Let us consider a uniaxial semiconductor slab to fill the space region $-l < z < l$; the space $|z| > 1$ is taken to be a vacuum. As in the case of a semi-infinite medium, the crystal is described by the dielectric permittivity tensor given in Eq. (5.1); the crystal axis lies in the xz-plane and makes an angle φ with the z-axis. Electromagnetic waves in the slab are a linear combination of the ordinary and extraordinary waves. The electric field vector of the surface wave can be written as

$$\boldsymbol{E}(\boldsymbol{r}, t) = \boldsymbol{E}(z)\exp\left[\mathrm{i}\left(k_x x + k_y y - \omega t\right)\right] \tag{6.1}$$

where

$$\boldsymbol{E}(z) = \boldsymbol{E}^{\mathrm{o}}_{+}\mathrm{e}^{a_{\mathrm{o}}z} + \boldsymbol{E}^{\mathrm{o}}_{-}\mathrm{e}^{-a_{\mathrm{o}}z} + \left(\boldsymbol{E}^{\mathrm{e}}_{+}\mathrm{e}^{a_{\mathrm{e}}z} + E^{\mathrm{e}}_{-}\mathrm{e}^{-a_{\mathrm{e}}z}\right)\mathrm{e}^{\mathrm{i}a_1 z}, \quad |z| < 1 \tag{6.1a}$$

$$\boldsymbol{E}(z) = \boldsymbol{E}_{+}\mathrm{e}^{-az}, \quad z > 1 \tag{6.1b}$$

$$\boldsymbol{E}(z) = \boldsymbol{E}_{-}\mathrm{e}^{az} \quad z < -1 \tag{6.1c}$$

Here a, a_{o}, a_1 and a_{e} are given by Eqs. (5.3), (5.8) and (5.9),

$$\boldsymbol{E}^{\mathrm{o}}_{\pm} = \mathrm{E}_{\pm}\left\{k_y\cos\varphi, (-k_x\cos\varphi \mp \mathrm{i}a_{\mathrm{o}}\sin\varphi), -k_y\sin\varphi\right\} \tag{6.2}$$

$$\boldsymbol{E}^{\mathrm{e}}_{\pm} = \left(\varepsilon_0\omega\varepsilon_{\perp}\varepsilon_{||}\right)^{-1}\left\{A_{\pm}\cos\varphi - B_{\pm}\sin\varphi, k_y\varepsilon_{||}\gamma_{\pm}, -(A_{\pm}\sin\varphi + B_{\pm}\cos\varphi)\right\} \tag{6.3}$$

$$A_{\pm} = \varepsilon_{||}\beta_{\pm}\gamma_{\pm} \qquad B_{\pm} = \varepsilon_{\perp}\left(k_y^2 + \beta_{\pm}^2\right) \tag{6.3a}$$

$$\beta_{\pm} = k_x\cos\varphi - (a_1 \mp \mathrm{i}a_{\mathrm{e}})\sin\varphi \qquad \gamma_{\pm} = k_x\sin\varphi + (a_1 \mp \mathrm{i}a_{\mathrm{e}})\cos\varphi \tag{6.4}$$

$$E_{\pm z} = \pm\frac{\mathrm{i}}{a}(k_x E_{\pm x} + k_y E_{\pm y}) \tag{6.5}$$

Radiowaves and Polaritons in Anisotropic Media. Roland H. Tarkhanyan and Nikolaos K. Uzunoglu

ISBN: 3-527-40615-8

The magnetic field associated with the surface wave is given by similar expressions, if one puts $\boldsymbol{H}^{\mathrm{o}}_{\pm}$, $\boldsymbol{H}^{\mathrm{e}}_{\pm}$ and $\boldsymbol{H}_{\pm}$ instead of $\boldsymbol{E}^{\mathrm{o}}_{\pm}$, $\boldsymbol{E}^{\mathrm{e}}_{\pm}$ and $\boldsymbol{E}_{\pm}$ in Eqs. (6.1), where

$$\boldsymbol{H}^{\mathrm{o}}_{\pm} = (\mu_0\omega)^{-1}E_{\pm}\left\{\left(k_x\rho_{\pm} - \frac{\omega^2}{c^2}\varepsilon_{\perp}\sin\varphi\right), k_y\rho_{\pm}, -\left(k_{\parallel}^2\cos\varphi \pm \mathrm{i}a_{\mathrm{o}}k_x\sin\varphi\right)\right\} \tag{6.6}$$

$$\rho_{\pm} = k_x\sin\varphi \mp \mathrm{i}a_{\mathrm{o}}\cos\varphi \tag{6.6a}$$

$$\boldsymbol{H}^{\mathrm{e}}_{\pm} = \boldsymbol{H}_{\pm}\left\{-k_y\cos\varphi, [k_x\cos\varphi - (a_1 \mp \mathrm{i}a_{\mathrm{e}})\sin\varphi], k_y\sin\varphi\right\} \tag{6.7}$$

$$\begin{aligned}\boldsymbol{H}_{\pm} &= (\mu_0\omega)^{-1}\left\{\pm\mathrm{i}a^{-1}\left[k_xk_yE_{\pm x} + \left(\frac{\omega^2}{c^2} - k_x^2\right)E_{\pm y}\right],\right.\\ &\quad \left.\pm\mathrm{i}a^{-1}\left[\left(k_y^2 - \frac{\omega^2}{c^2}\right)E_{\pm x} - k_xk_yE_{\pm y}\right], \left(k_xE\pm y - k_yE_{\pm x}\right)\right\},\end{aligned} \tag{6.8}$$

Continuity of tangential components of the fields $\boldsymbol{E}$ and $\boldsymbol{H}$ at $z = +l$ and $z = -l$ yields a set of eight linear homogeneous algebraic equations for eight unknowns $E_{\pm}$, $H_{\pm}$, $E_{\pm x}$ and $E_{\pm y}$. Nontrivial solutions are obtained by setting the determinant of coefficients equal to zero. This condition gives ω as a function of k_x and k_y for an arbitrary configuration of the crystal axis, surfaces of the slab and the wave vector $\boldsymbol{k}_{\parallel} = \{k_x, k_y\}$ which is parallel to the surfaces. The general dispersion relation has the form

$$\boldsymbol{G}(\omega, k_x, k_y, \varphi) = 0 \tag{6.9}$$

where $\boldsymbol{G}$ is a cumbersome, lengthy and unwieldy nonlinear function. We do not give here the explicit form of the function $\boldsymbol{G}$ in Eq. (6.9), resulting from the boundary conditions, and restrict ourselves to several specific cases for the orientation of the crystal axis relative to the surface and the wave vector.

A. The crystal axis is perpendicular to the surface ($\varphi = 0$)
B. $\varphi = \pi/2$ and $\boldsymbol{k}_{\parallel}$ is parallel to the crystal axis ($k_y = 0$)
C. $\varphi = \pi/2$ and $\boldsymbol{k}_{\parallel}$ is perpendicular to the crystal axis ($k_x = 0$)

The surface wave is an extraordinary one in cases A and B, and an ordinary one in the case C. For all these three cases the magnetic field of the wave is parallel to the surface and perpendicular to $\boldsymbol{k}_{\parallel}$, that is the wave is of TM type. In addition, for all these cases the dispersion relation (6.9) splits in two, one of which corresponds to the symmetric and the second to the antisymmetric wave with respect to the central plane $z = 0$:

$$\text{A.}\ \tanh(la_{\mathrm{e}}) = -\frac{a\varepsilon_{\perp}}{a_{\mathrm{e}}} \tag{6.10a}$$

$$\coth(la_{\mathrm{e}}) = -\frac{a\varepsilon_{\perp}}{a_{\mathrm{e}}} \tag{6.10b}$$

B. $\tanh(la_e) = -\frac{a\varepsilon_{||}}{a_e}$ (6.11a)

$$\coth(la_e) = -\frac{a\varepsilon_{||}}{a_e} \tag{6.11b}$$

C. $\tanh(la_o) = -\frac{a\varepsilon_{\perp}}{a_o}$ (6.12a)

$$\coth(la_o) = -\frac{a\varepsilon_{\perp}}{a_o} \tag{6.12b}$$

The magnetic field is symmetrical $[H \sim \cosh(a_e z)]$ for the waves described by Eqs. (6.10a), (6.11a) and (6.12a), while it is antisymmetrical $[H \sim \sinh(a_e z)]$ for the waves (6.10b), (6.11b) and (6.12b). Nonradiative surface waves exist only if $\varepsilon_{\perp} < 0$ for the cases A and C, and if $\varepsilon_{||} < 0$, for the case B. The decay constants a, a_o and a_e must be real and positive.

6.2
Surface Polaritons in a Polar Semiconductor Slab

Let us consider now the surface waves in a uniaxial polar semiconductor slab, where the coupling of optical phonons to plasmons leads to an interesting behavior of the surface polaritons, which differ in certain significant details from those in a cubic (isotropic) polar semiconductor slab as well as in a uniaxial semi-infinite semiconductor.

In Fig. 6.1, the regions of existence of the surface phonon–plasmon polaritons are shown for the case $\omega_{T||} < \omega_{T\perp} < \omega_{L||} < \omega_{L\perp}$ (CdS). In each region there is a pair of symmetrical and antisymmetrical modes. In contrast to the isotropic case, the number of slow modes for which the phase velocity $\omega/k_{||} \to 0$, as well as the boundaries of the existence regions, depends on the carrier density, orientation of the crystal axis relative to the surface and the direction of propagation.

In the case A (see previous section) the surface waves exist in four regions R_1 to R_4 (Fig. 6.1a). The regions in which the surface modes exist over a finite range in $k_{||}$ represent special interest because for an isotropic slab such modes do not occur. There are two such regions (R_2 and R_4, Fig. 6.1a) if

$$\omega_{0\perp} < \omega_{T||}\sqrt{\frac{\omega_{L\perp}^2 - \omega_{T||}^2}{\omega_{T\perp}^2 - \omega_{T||}^2} - \frac{1}{\varepsilon_{\perp\infty}}} \equiv \omega_1 \tag{6.13}$$

and only a single one if $\omega_{0\perp} > \omega_1$ (R_4, Fig. 6.1b); in the last case the region R_2 is replaced by the region $R_6 + R_7$. The regions R_2, R_4 and $R_6 + R_7$ are bounded by the light line $\omega = k_{||}c$ and by the curves

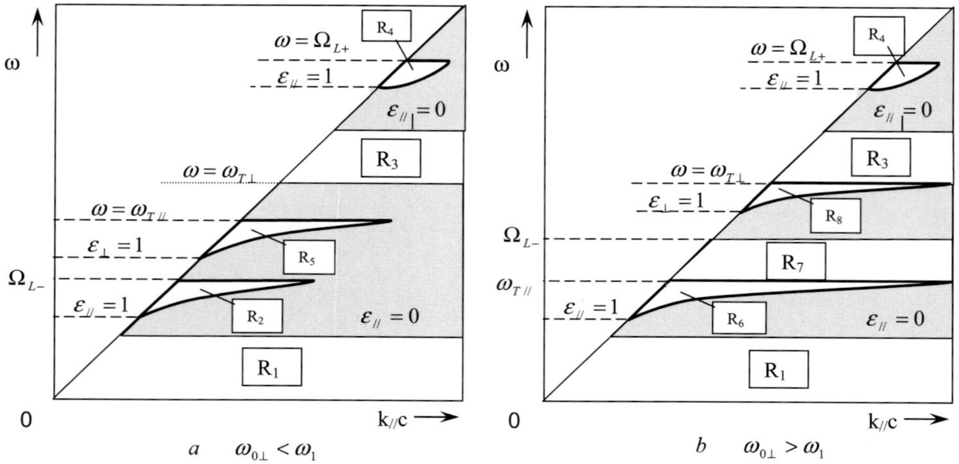

Fig. 6.1 Regions of existence of the surface phonon–plasmon polaritons. The shaded regions and those to the left of the light line $\omega = ck_{||}$ are those in which the surface waves cannot propagate.

$$k_{||}c = \omega\sqrt{\varepsilon_{||}(\omega)} \tag{6.14}$$

and $\varepsilon_{\perp}(\omega) = 0$ where $\varepsilon_{||,\perp}(\omega)$ are given by Eqs. (2.19). The solutions of Eq. (6.14) coincide with the frequencies of the longitudinal phonon–plasmon branches $\Omega_{L\pm}$ given by Eq. (2.22). Both Ω_{L+} and Ω_{L-} increase with increasing $N \sim \omega_{0\perp}^2$. Note that $\Omega_{L-} = \omega_{T||}$ if $\omega_{0\perp} = \omega_1$.

In the case B, the surface waves exist in three regions R_1, R_3 and R_5 (Fig. 6.1a) if

$$\omega_{0\perp} < \omega_{T||}\sqrt{\frac{\omega_{L\perp}^2 - \omega_{T||}^2}{\omega_{T\perp}^2 - \omega_{T||}^2} - \frac{1}{\varepsilon_{\perp\infty}}} \equiv \omega_2 \tag{6.15}$$

and only in the regions R_1 and R_3 if $\omega_2 < \omega_{0\perp} < \omega_1$. For $\omega_{0\perp} > \omega_1$, there are three regions again: R_1, R_7 and $R_3 + R_8$ (Fig. 6.1b). The regions R_5 and R_8 lie above and to the left of the curve $k_{||}c = \omega\sqrt{\varepsilon_{\perp}(\omega)}$. The value of the frequency at which this curve intersects the light line, that is the smaller solution of the equation

$$\varepsilon_{\perp}(\omega) = 1 \tag{6.16}$$

coincides with $\omega_{T||}$ if $\omega_{0\perp} = \omega_2$.

In the case C, the surface polaritons exist, as in the case of an isotropic slab, only in frequency regions

$$\omega < \Omega_{L-} \quad \omega_{T\perp} < \omega < \Omega_{L+} \tag{6.17}$$

lying to the right of the light line.

6.3 Quasielectrostatic Surface Waves

Let us consider now the dispersion curves in the nonretardation limit $\omega/k_{||} << c$ Then, the macroscopic electric field associated with the surface wave can be written in the form

$$E = -grad\psi \quad \psi(r, t) = \varphi(z)\exp\left[i\left(k_x x + k_y y - \omega t\right)\right] \tag{6.18}$$

where

$$\varphi(z) = \left(\varphi_+ e^{-\beta z} + \varphi_- e^{\beta z}\right)e^{ia_1 z} \quad |z| < l \tag{6.19a}$$

$$= \varphi_0 e^{-k_{||}z}, \qquad z > l \tag{6.19b}$$

$$= \varphi_1 e^{k_{||}z} \qquad z < -l \tag{6.19c}$$

$$\beta = |\varepsilon_{zz}|^{-1}\sqrt{\varepsilon_\perp\left(\varepsilon_{||}k_x^2 + \varepsilon_{zz}k_y^2\right)} \tag{6.20}$$

and we require β to be real. Using continuity conditions for the tangential components of the vector $\boldsymbol{E}$ and for the normal component of the vector $\boldsymbol{D} = \varepsilon_0\varepsilon\boldsymbol{E}$ *at* $z = \pm l$, after eliminating φ_0 and φ_1 we obtain

$$\left(k_{||} - \varepsilon_{zz}\beta\right)\varphi_+ e^{-\beta l} + \left(k_{||} + \varepsilon_{zz}\beta\right)\varphi_- e^{\beta l} = 0 \tag{6.21a}$$

$$\left(k_{||} + \varepsilon_{zz}\beta\right)\varphi_+ e^{-\beta l} + \left(k_{||} - \varepsilon_{zz}\beta\right)\varphi_- e^{\beta l} = 0 \tag{6.21b}$$

The condition of solvability of the system (6.21a) and (6.21b) gives

$$\left(k_{||} - \varepsilon_{zz}\beta\right)^2 e^{-2\beta l} - \left(k_{||} + \varepsilon_{zz}\beta\right)^2 e^{2\beta l} = 0 \tag{6.22}$$

which leads to two dispersion relations in the form

$$\tanh(\beta l) = -\frac{\varepsilon_{zz}\beta}{k_{||}} \tag{6.23}$$

$$\coth(\beta l) = -\frac{\varepsilon_{zz}\beta}{k_{||}} \tag{6.24}$$

Equation (6.23) corresponds to the antisymmetric $\left(\varphi_+ = -\varphi_-\right)$ and Eq. (6.24) to the symmetric $\left(\varphi_+ = \varphi_-\right)$modes. Note that for the specific cases A, B, C, the relations (6.23) and (6.24) can be obtained immediately by formally taking the limit $c \to \infty$ in Eqs. (6.10a)-(6.12b). It can easily be seen that for real and positive β, the surface waves exist only if $\varepsilon_{zz} < 0$. For a given value of the angle φ between the crystal axis and the normal to the surface of the slab, these conditions may be rewritten in the equivalent form

$$\varepsilon_{\perp}(\omega) < 0, \quad \varepsilon_{\|}(\omega) < \frac{|\varepsilon_{\perp}(\omega)|\tan^2\varphi \cdot \tan^2\eta}{1 + \tan^2\varphi + \tan^2\eta} \tag{6.25}$$

where $\tan\eta = k_y/k_x$. Using Eqs. (5.1a), (6.20), (6.23) and (6.24), we can determine the dependence of the coupled nonretarded surface modes' frequencies on the dimensionless quantity $k_{\|}l$. For a thin film ($l\beta \ll 1$) Eqs. (6.23) and (6.24) give

$$\varepsilon_{zz} = -k_{\|}l \tag{6.26}$$

$$\varepsilon_{zz} = -\varepsilon_{\perp}k_{\|}l(\varepsilon_{\|}\cos^2\eta + \varepsilon_{zz}\sin^2\eta) \tag{6.27}$$

We see that the frequencies of the antisymmetrical modes described by Eq. (6.26) are independent of the direction of propagation, and those for the symmetrical modes are independent of the crystal-axis orientation if $\eta = \pi/2$. The frequency of the antisymmetrical mode decreases with increasing $k_{\|}l$ while that of the symmetrical mode increases. As $k_{\|}l \to \infty$ asymptotic values of ω approach the frequencies for nonretarded surface waves at the interface between a vacuum and a semi-infinite uniaxial crystal, which are defined by the equation

$$\varepsilon_{\perp}(\varepsilon_{\|}\cos^2\eta + \varepsilon_{zz}\sin^2\eta) = 1 \tag{6.28}$$

The number of solutions of Eqs. (6.23) and (6.24) changes in dependence on the parameters φ, η and N. In case A (see Section 6.1) there are two pairs of surface modes which exist in the frequency regions

$$\omega < \omega_{L-} \quad \text{and} \quad \omega_{T\perp} < \omega < \omega_{L+} \tag{6.29}$$

if $\omega_{0\perp} < \omega_1$. Here ω_1 is given by Eq. (6.13) and $\omega_{L\pm}$ are defined by Eq. (2.21). These modes are shown in Fig. 6.2 (solid lines). The high-frequency curves correspond to the surface phonon-like modes and the low-frequency ones to the plasmon-like modes. An additional pair of the coupled phonon–plasmon modes appears in the region

$$\omega_{T\|} < \omega < \Omega_{L-} \tag{6.30}$$

when $\omega_{0\perp} > \omega_1$. These new modes are a result of the crystal anisotropy. In contrast to the phonon-like and plasmon-like modes, they do not exist for the values

$$k_{\|}l < |\varepsilon_{\|}(\Omega_{L-})| \tag{6.31}$$

(upper curve) and

$$k_{\|}l < \left|\varepsilon_{\perp}\left(\omega_{T\|}\right)\right|^{-1} \tag{6.32}$$

(lower curve). Thus, there is a minimum value for the slab thickness (at a given value of $k_{\|}$) and a maximum value for the surface wavelength (at a given thickness) for the propagation of the modes.

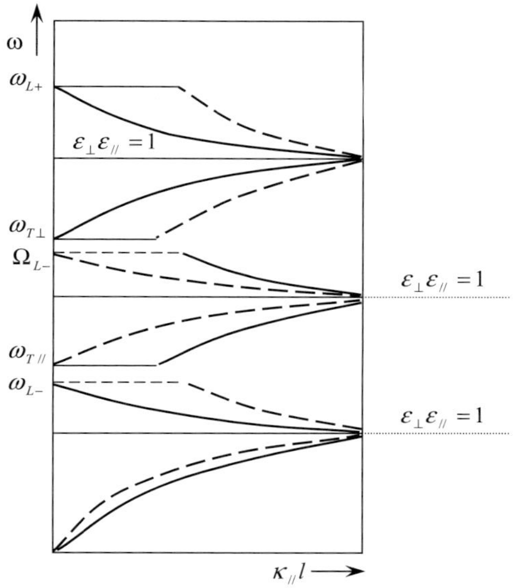

Fig. 6.2 A schematic plot of ω versus $k_{\|}l$ for the nonretarded surface modes. Solid curves correspond to the case A and dashed curves to the case B.

Dispersion curves in case B are given in Fig. 6.2 by dotted lines. For $\omega_{0\perp} < \omega_1$ there are phonon-like modes in the region

$$\omega_{T\perp} < \omega < \omega_{L\perp} \tag{6.33}$$

and plasmon-like modes below $\omega = \omega_{L-}$. The lower plasmon-like branch exists for all values of $k_{\|}l$ while the upper branch exists only for

$$k_{\|}l > |\varepsilon_{\perp}(\omega_{L-})| \tag{6.34}$$

The upper phonon-like branch exists for

$$k_{\|}l > |\varepsilon_{\perp}(\omega_{L+})| \tag{6.35}$$

and the lower one for

$$k_{\|}l > \frac{1}{|\varepsilon_{\|}(\omega_{T\perp})|} \tag{6.36}$$

If $\omega_{0\perp} > \omega_1$, a pair of new coupled surface modes appears, which exist in the frequency region

$$\omega_{T||} < \omega < \Omega_{L-} \tag{6.37}$$

for all values of $k_{||}l$ and which are absent in an isotropic medium. In the limit $k_{||}l \to \infty$, the frequencies of all unretarded surface modes approach the asymptotic values defined by the equation

$$\varepsilon_\perp(\omega)\varepsilon_{||}(\omega) = 1 \tag{6.38}$$

In the case C there are no additional modes due to the crystal anisotropy. Two symmetrical and two antisymmetrical waves exist. The frequencies of the symmetrical modes rise from $\omega = 0$ and $\omega = \omega_{T\perp}$ at $k_{||} = 0$ to the asymptotic values given by

$$\omega^2 = \frac{1}{2}\left(1 + \frac{1}{\varepsilon_{\perp\infty}}\right)^{-1} \times \left\{\omega_{0\perp}^2 + \omega_{L\perp}^2 + \frac{\omega_{T\perp}^2}{\varepsilon_{\perp\infty}} \pm \left[\left(\omega_{0\perp}^2 + \omega_{L\perp}^2 + \frac{\omega_{T\perp}^2}{\varepsilon_{\perp\infty}}\right)^2 - 4\left(1 + \frac{1}{\varepsilon_{\perp\infty}}\right)\omega_{0\perp}^2\omega_{T\perp}^2\right]^{1/2}\right\} \tag{6.39}$$

The frequencies of the antisymmetrical waves decrease with increasing $k_{||}l$ from Ω_{L+} to the values (6.39).

Finally, let us consider briefly one more interesting effect due to the crystal anisotropy. One can easily find that in an isotropic semiconductor slab there are sinusoidal bulk modes with exponentially decaying fields outside the slab, but such guide modes occur only when the retardation effect is included. Crystal anisotropy leads to the possibility of propagation for nonretarded sinusoidal modes, which are described by the dispersion relations

$$\tan(l\beta') = -\frac{\beta'\varepsilon_{zz}}{k_{||}} \tag{6.40a}$$

$$\cot(l\beta') = \frac{\beta'\varepsilon_{zz}}{k_{||}} \tag{6.40b}$$

where $\beta' = -i\beta$. These modes exist only if β' is real. Equations (6.40a) and (6.40b) correspond to the solutions of even parity [$\varphi \sim \cos(\beta' z)$] and odd parity [$\varphi \sim \sin(\beta' z)$], respectively. For an n-type CdS crystal slab, in the cases A and B there are three frequency regions in which the sinusoidal modes can occur:

$$\omega_{L-} < \omega < \omega_{T||}, \quad \Omega_{L-} < \omega < \omega_{T\perp}, \quad \omega_{L+} < \omega < \Omega_{L+} \tag{6.41}$$

For the case C one has $\beta = k_{||}$ and, as in an isotropic material, there are no nonretarded oscillatory solutions.

6.4
Influence of an External Magnetic Field

Let us consider as an another example of an anisotropic medium a semiconductor slab in an external static magnetic field $\boldsymbol{B}_0$, which is taken to lie in the plane parallel to the surfaces. Surface magnetoplasmons in a semi-infinite medium have been considered by Chiu and Quinn [62] and Wallis et al. [47]. An interesting result was found for $\boldsymbol{k}_{||} \perp \boldsymbol{B}_0$: nonreciprocity between $-\boldsymbol{k}_{||}$ and $+\boldsymbol{k}_{||}$. As we will see, this is not true for a slab of finite thickness; the presence of the second face essentially changes the properties of the surface magnetoplasmons [51].

Let the coordinate system be chosen in such a way that the x-axis is normal to the surfaces and the magnetic field $\boldsymbol{B}_0$ is parallel to the z-axis. The semiconductor slab will be assumed to occupy the space region $-l < x < l$. The dielectric permittivity tensor can be written in the form

$$\varepsilon = \begin{pmatrix} \varepsilon_1 & \mathrm{i}\varepsilon_2 & 0 \\ -\mathrm{i}\varepsilon_2 & \varepsilon_1 & 0 \\ 0 & 0 & \varepsilon_3 \end{pmatrix} \tag{6.42}$$

For an n-type semiconductor with isotropic effective mass m of the conduction electrons (for example, n-InSb), the components of the permittivity tensor are expressed by

$$\varepsilon_1 = \varepsilon_\infty \left(1 - \frac{\omega_\mathrm{p}^2}{\omega^2 - \omega_\mathrm{c}^2}\right) \tag{6.43}$$

$$\varepsilon_2 = -\frac{\varepsilon_\infty\, \omega_\mathrm{c}\omega_\mathrm{p}^2}{\omega\left(\omega^2 - \omega_\mathrm{c}^2\right)}, \quad \varepsilon_3 = \varepsilon_\infty \left(1 - \frac{\omega_\mathrm{p}^2}{\omega^2}\right) \tag{6.44}$$

where

$$\omega_\mathrm{p}^2 = \frac{Ne^2}{m\varepsilon_0\varepsilon_\infty}, \quad \omega_\mathrm{c} = \frac{eB_0}{m} \tag{6.45}$$

so that ω_p is the plasma frequency and ω_c is the cyclotron frequency.

We assume that both the slab thickness $2l$ and the length of the considered waves are large compared to the lattice spacing and the cyclotron radius $R = (\hbar/eB_0)^{1/2}$ We will only consider the situation when the direction of propagation is perpendicular to the magnetic field: $k_z = 0$, $k_y = k_{||}$. Using the Maxwell's equations and standard boundary conditions on the planes $x = \pm l$, one can easily show that there are no TE-type surface waves with $\boldsymbol{E}\|\boldsymbol{B}_0$. The fields associated with TM-type surface waves are given by

$$\boldsymbol{E} = \left(E_x, E_y, 0\right) \qquad \boldsymbol{H} = (0, 0, H(x))\exp\left[i(k_{\|}y - \omega t)\right] \tag{6.46}$$

where

$$E_x = -\frac{k_{\|}}{\varepsilon_0\omega}H_z \qquad E_y = -\frac{i}{\varepsilon_0\omega}\frac{dH_z}{dx} \quad \text{for} \quad |x| > 1 \tag{6.47a}$$

$$E_x = -\frac{1}{\varepsilon_0\omega\varepsilon_v}\left(k_{\|}H_z + \frac{\varepsilon_2}{\varepsilon_1}\frac{dH_z}{dx}\right), E_y = -\frac{i}{\varepsilon_0\omega\varepsilon_v}\left(\frac{k_{\|}\varepsilon_2}{\varepsilon_1}H_z + \frac{dH_z}{dx}\right) \quad \text{for} \quad |x| < 1 \tag{6.47b}$$

$$H(x) = H_+e^{-ax} \quad x > 1 \tag{6.48a}$$

$$H(x) = H_1e^{-a_vx} + H_2e^{a_vx} \qquad |x| < 1 \tag{6.48b}$$

$$H(x) = H_-e^{ax} \quad x < -1 \tag{6.48c}$$

$$a = \sqrt{k_{\|}^2 - \frac{\omega^2}{c^2}} \qquad a_v = \sqrt{k_{\|}^2 - \varepsilon_v\frac{\omega^2}{c^2}} \tag{6.49}$$

$$\varepsilon_v \equiv \varepsilon_1 - \frac{\varepsilon_2^2}{\varepsilon_1} \tag{6.50}$$

Using the continuity conditions for E_y and H_z at the planes $x = \pm\, l$, we obtain a dispersion relation for the waves in the form

$$\tanh(a_vL) = -\frac{2aa_v}{a^2 + a_v^2 + \varepsilon k_{\|}^2} \tag{6.51}$$

where

$$\varepsilon \equiv \varepsilon_v - 2 + \varepsilon_1^{-1} \tag{6.52}$$

Solutions of Eq. (6.51) give ω as a function of $k_{\|}$ for given values of $\boldsymbol{B}_0$ and slab thickness $L = 2l$. They correspond to the nonradiative surface waves only if the following conditions are fulfilled:

$$\left(\frac{ck_{\|}}{\omega}\right)^2 > \max\{\varepsilon_v, 1\} \tag{6.53}$$

$$\varepsilon < 0 \tag{6.54}$$

$$\left(\varepsilon_v + \varepsilon_1^{-1}\right)\left(\frac{ck_{\|}}{\omega}\right)^2 < \varepsilon_v + 1 \tag{6.55}$$

The first condition ensures that the energy stored in the surface wave cannot radiate into the vacuum as well as into the slab, since no bulk modes occur when a and a_v are real and positive. The second condition lays down some restrictions on the value of the magnetic field. To see this, let us consider a definite value of ω from the frequency region $\omega \leq \omega_p$. Then, the condition (6.54) is fulfilled only for

$$\omega_c < \omega_{c0} \equiv \left| \omega - \frac{\varepsilon_\infty \omega_p^2}{(\varepsilon_\infty - 1)\omega} \right| \tag{6.56}$$

The corresponding maximum value of B_0 decreases with increasing ω. Conditions (6.53) and (6.55) lead to additional restrictions on the value of B_0 as well as $k_\parallel$. In Fig. 6.3, the region in the plane $\{ck_\parallel/\omega, \omega_c\}$ is shown for which the conditions (6.53)–(6.55) are fulfilled when $\omega < \omega_p$. On the right-hand side, this region is bounded by curves defined by the equations

$$\frac{c^2 k_\parallel^2}{\omega^2} = \frac{\varepsilon_\nu + 1}{\varepsilon_\nu + \varepsilon_1^{-1}} \tag{6.57}$$

(upper curve) and

$$\frac{c^2 k_\parallel^2}{\omega^2} = \varepsilon_\nu \tag{6.58}$$

(lower curve). The value of ω_c at which these curves are ω intersected corresponds to the upper limit of B_0, for which the propagation of the surface waves is yet possible. This value is specified by

$$\omega_{c1} = \frac{\omega_p}{\sqrt{2\rho(1-\varsigma)}} \left\{ \varsigma + 2\rho\varsigma^2 + \frac{1}{\varepsilon_\infty} + \frac{1}{\sqrt{\varepsilon_\infty}} \left[\rho\left(5\varsigma^2 - \frac{1}{\varepsilon_\infty}\right) + (\varsigma+1)^2 \right]^{1/2} \right\}^{1/2} \tag{6.59}$$

where $\varsigma \equiv 1 - (\omega/\omega_p)^2$ and $\rho \equiv \varepsilon_\infty - 1$. If $\omega < \omega_p$ one has $\omega_{c1} < \omega_{c0}$ that is, the upper limit of B_0 is smaller than that required by the condition (6.54). Note that for

$$\omega_c > \omega_{c2} \equiv \frac{\omega_p}{\sqrt{1+\varepsilon_\infty^2}} \left(1 - \varsigma + \frac{\varepsilon_\infty^2 \varsigma^2}{1-\varsigma} \right)^{1/2} \tag{6.60}$$

the surface waves only exist in a finite range of $k_\parallel$ and the upper limit of $k_\parallel$ decreases monotonically with increasing B_0. For $\omega_c > \omega_{c3}$ where

$$\omega_{c3} = \omega_p \sqrt{\frac{\varsigma(\varsigma\rho + 1)}{\rho(1-\varsigma)}} \tag{6.61}$$

the lower limit of $k_{||}$ is shifted towards the shorter-wavelength side with increasing B_0; then the range of $k_{||}$ shortens and vanishes at $\omega_c = \omega_{c1}$.

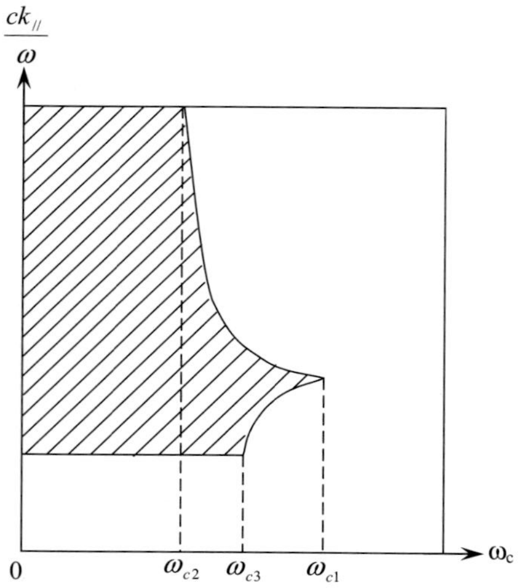

Fig. 6.3 The existence region for the surface waves in the plane $(ck_{||}/\omega, \omega_c)$.

Before we proceed with the discussion of Eq. (6.51), one important point must be mentioned. For a semi-infinite medium the surface waves are described by the dispersion relation

$$a_\nu + a\varepsilon_\nu = \frac{\varepsilon_2}{\varepsilon_1} k_{||} \tag{6.62}$$

which can be obtained immediately from Eq. (6.51), if one rewrites the latter in the form

$$\tanh(a_\nu L) = -\frac{2aa_\nu\varepsilon_\nu}{a_\nu^2 + a^2\varepsilon_\nu^2 - k_{||}^2(\varepsilon_2/\varepsilon_1)^2} \tag{6.63}$$

and takes the limit $L \to \infty$ As follows from Eq. (6.62), the positive and negative values of $\boldsymbol{k}_{||}$ are not equivalent. Unlike this, $k_{||}$ appears only as even powers in Eq. (6.51), so that the reciprocity of $+\boldsymbol{k}_{||}$ and $-\boldsymbol{k}_{||}$ is rehabilitated, which is a result of the finite thickness of the slab.

For a "thick" slab ($a_\nu L >> 1$) Eq. (6.51) gives

$$a + a_\nu = |k_{||}|\sqrt{-\varepsilon} \tag{6.64}$$

Using Eq. (6.64), we obtain

$$\frac{c^2 k_{||}^2}{\omega^2} = \frac{1 + \varepsilon_v \pm 2\varepsilon_2 \varepsilon_1^{-1} |\varepsilon|^{-1/2}}{4 + \varepsilon} \tag{6.65}$$

which corresponds to two surface waves with the same direction of propagation, but with different phase velocities. The wave localized at the given face of a thick slab is identical with the wave localized at the opposite face and propagating in the opposite direction. Note that Eq. (6.64) can also be obtained from Eq. (6.62) describing the surface waves in a semi-infinite case. However, in the latter case, in contrast to the case of the slab with finite thickness, only one from two waves described by Eq. (6.64) can propagate along the given direction perpendicular to B_0.

The frequencies

$$\omega = \sqrt{\frac{\omega_c^2}{4} + \frac{\varepsilon_\infty \omega_p^2}{\varepsilon_\infty + 1}} \pm \frac{\omega_c}{2} \tag{6.65a}$$

and

$$\omega = \sqrt{\frac{\omega_c^2}{4} + \frac{\varepsilon_\infty \omega_p^2}{\varepsilon_\infty - 1}} \pm \frac{\omega_c}{2} \tag{6.65b}$$

correspond to the resonances $(k_{||} = 0)$, and the frequencies

$$\omega^2 = \frac{1}{2}\left[A \pm \sqrt{A^2 - \frac{4\varepsilon_\infty \omega_p^4}{\varepsilon_\infty - 1}}\right], \qquad A \equiv \omega_c^2 + \frac{(2\varepsilon_\infty - 1)\omega_p^2}{\varepsilon_\infty - 1} \tag{6.66}$$

correspond to cut-off $(k_{||} = 0)$ for the waves given by Eq. (6.64).

For a "thin" slab $(a_v L << 1)$ Eq. (6.51) gives

$$\frac{c^2 k_{||}^2}{\omega^2} = \left(\varepsilon_v + \varepsilon_1^{-1}\right)^2 \left[2a + (\varepsilon_v + 1)\left(\varepsilon_v + \varepsilon_1^{-1}\right) \pm 2\sqrt{a^2 + a\left(\varepsilon_v + \varepsilon_1^{-1}\right)\left(1 - \varepsilon_1^{-1}\right)}\right] \tag{6.67}$$

where $a = (c/L\omega)^2$ Equation (6.67) corresponds to two surface waves with the same frequency and propagation direction, but with different wavelengths, as in the case of a thick slab. The phase velocity of the wave with the + sign in Eq. (6.67) becomes zero at the resonance frequencies given by

$$\omega^2 = \frac{\omega_c^2}{2} + \frac{\omega_p^2}{1 + \varepsilon_\infty^{-2}} \pm \sqrt{\frac{\omega_c^4}{4} + \frac{\omega_p}{1 + \varepsilon_\infty^{-2}}\left(\omega_c^2 - \frac{\omega_p^2}{1 + \varepsilon_\infty^2}\right)} \tag{6.68}$$

For the second wave, the phase velocity is finite at these frequencies and is equal to

$$v_{\mathrm{p}} = \frac{c}{\sqrt{1+\frac{(1+\varepsilon_\nu)^2}{4a}}} \tag{6.69}$$

Note that for $\omega_{\mathrm{c}} < \omega_{\mathrm{c}2}$ and $\omega < \omega_{\mathrm{p}}$, the waves described by Eq. (6.67) exist only if

$$\frac{L\omega}{c} < \left[\left(\varepsilon_\nu + \varepsilon_1^{-1}\right)\left(\varepsilon_1^{-1} - 1\right)\right]^{-1/2} \tag{6.70}$$

In the case of the reverse inequality $k_{||}$ has an imaginary part, and the waves will attenuate when they propagate parallel to the surface.

Finally, let us consider briefly the surface waves in the absence of retardation, that is in the limit $c \to \infty$ In this case Eq. (6.51) gives

$$\tanh(Lk_{||}) = -\frac{2\varepsilon_1}{1+\varepsilon_1^2 - \varepsilon_2^2} \tag{6.71}$$

from which we obtain

$$\omega^2 = \frac{1}{2}\left\{\omega_{\mathrm{c}}^2 + f\left(k_{||}\right)\omega_{\mathrm{p}}^2 \pm \left[\left(\omega_{\mathrm{c}}^2 + f\left(k_{||}\right)\omega_{\mathrm{p}}^2\right)^2 - \frac{2\varepsilon_\infty f\left(k_{||}\right)\omega_{\mathrm{p}}^4}{\varepsilon_\infty + \coth\left(Lk_{||}\right)}\right]^{1/2}\right\} \tag{6.72}$$

where

$$f(k_{||}) = \frac{\varepsilon_\infty + \coth\left(Lk_{||}\right)}{\frac{1}{2}\left(\varepsilon_\infty + \varepsilon_\infty^{-1}\right) + \coth\left(Lk_{||}\right)} \tag{6.72a}$$

Equation (6.72) corresponds to two branches of unretarded surface magnetoplasmons. The frequency of the lower branch increases with increasing $Lk_{||}$ from the origin to the asymptotic value given by

$$\omega^2 = \frac{\omega_{\mathrm{c}}^2}{2} + \frac{\omega_{\mathrm{p}}^2}{1+\varepsilon_\infty^{-1}} - \omega_{\mathrm{c}}\sqrt{\frac{\omega_{\mathrm{c}}^2}{4} + \frac{\omega_{\mathrm{p}}^2}{1+\varepsilon_\infty^{-1}}} \tag{6.73}$$

The frequency of the upper branch decreases or increases with increasing $Lk_{||}$ from the value

$$\omega = \sqrt{\omega_{\mathrm{c}}^2 + \omega_{\mathrm{p}}^2} \tag{6.74}$$

at $k_{||} = 0$ to the asymptotic value which differs from that defined in Eq. (6.73) only by the opposite sign in front of the square root. It decreases if

$$\omega_{\mathrm{c}} < \varpi_{\mathrm{c}} \equiv \frac{\omega_{\mathrm{p}}}{\sqrt{\varepsilon_\infty^2 - 1}} \tag{6.75}$$

and increases if $\omega_c > \varpi_c$. For a given frequency in the region $\omega \le \omega_p$, unretarded surface magnetoplasmons exist only if the following conditions are fulfilled:

$$\omega_c < \omega_{c4} \equiv \omega \frac{|1+\varepsilon_3|}{1+\varepsilon_\infty} \tag{6.76}$$

$$\tanh(Lk_{||}) > -\frac{2\varepsilon_3}{1+\varepsilon_3^2} \tag{6.77}$$

Since $\omega_{c4} < \omega_{c2}$ we conclude that the upper limit of the static magnetic field for which the surface waves exist increases when the retardation is included. As follows from Eq. (6.77), the minimal allowed value of the parameter $Lk_{||}$ is a function of the free carrier density N: it increases with increasing N, arrives at a maximum value for

$$\omega_p^2 = \omega^2(1+\varepsilon_\infty^{-1}) \tag{6.78}$$

and then decreases. Note that $\omega_{c4} = 0$ when the condition (6.78) is fulfilled, and there are no surface magnetoplasmons in this case. In the region $\omega > \omega_p$, they exist for all values of $Lk_{||}$, but only in a limited region of B_0 between $\omega_c = \sqrt{\omega^2 - \omega_p^2}$ and $\omega_c = \omega_{c4}$

As to the dependence on B_0 of the frequencies given by Eq. (6.72), it is shown schematically in Fig. 6.4. We see that for a given value of $Lk_{||}$, they shift to opposite sides: the frequency of the upper branch increases monotonically with increasing ω_c and approaches asymptotically the line $\omega = \omega_c$, while for the lower branch it decreases and tends to zero.

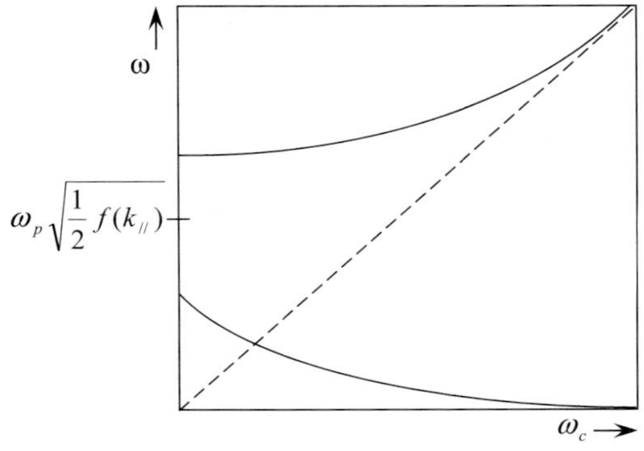

Fig. 6.4 A schematic plot of ω versus ω_c for the nonretarded surface magnetoplasmons.

7
Interface Magnon–Plasmon Polaritons and Total Transmission of Electromagnetic Waves Through a Semiconductor/Antiferromagnet Layered Structure

7.1
Dispersion Relations and Conditions Necessary for the Existence of Interface Magnon–Polaritons

There is a profound difference between the properties of surface polaritons in nonmagnetic and magnetic crystals. As has been mentioned in the Introduction, only TM-type surface waves exist between two nonmagnetic media while only TE-type modes can travel along an interface between a magnetic and a nonmagnetic insulator (see also Ref. [62]). Interface polaritons in a layered structure consisting of a magnetic material and a nonmagnetic semiconductor have been first studied in Ref. [58]. The main reasons why the interface between a magnetic insulator and a semiconductor (e.g. between $Cd_{1-x}Mn_xTe$ and CdTe) is of interest are the following. First, the polaritons travelling along such an interface are a peculiar "mixture" of the waves belonging to three subsystems: plasmons, optical magnons and "pure" electromagnetic waves localized near the interface. Secondly, both TM- and TE-type coupled waves can propagate along such interfaces, as we will see soon.

Consider an antiferromagnetic crystal with two antiparallel magnetic sublattices. Let us assume that a uniaxial antiferromagnet occupies the space region $z<0$ and has an "easy-axis" magnetic anisotropy so that the saturation magnetization $\boldsymbol{M}_s$ of each sublattice is parallel to the optical axis $\boldsymbol{C}$ and the anisotropy field $\boldsymbol{H}_A$ as well as the exchange field $\boldsymbol{H}_e$ are oriented in the same direction. An isotropic nonpolar semiconductor occupies the region $z>0$ and contains only one type of free charge carriers; its high-frequency properties are characterized by the collisionless scalar permittivity

$$\varepsilon(\omega) = \varepsilon_\infty \left(1 - \frac{\omega_p^2}{\omega^2}\right), \tag{7.1}$$

where ω_p is the plasma frequency defined by Eq. (6.45).

Let us first consider the case when the optical axis $\boldsymbol{C}$ of the antiferromagnet is perpendicular to the interface: C||oz. Then the magnetic permeability tensor is given by [73]

Radiowaves and Polaritons in Anisotropic Media. Roland H. Tarkhanyan and Nikolaos K. Uzunoglu

ISBN: 3-527-40615-8

$$\mu = \begin{pmatrix} \mu_\perp & 0 & 0 \\ 0 & \mu_\perp & 0 \\ 0 & 0 & 1 \end{pmatrix} \tag{7.2}$$

where

$$\mu_\perp = \frac{\omega^2 - \omega_L^2}{\omega^2 - \omega_T^2} \tag{7.3}$$

Here

$$\omega_T = \gamma\sqrt{2H_A H_s} \tag{7.4}$$

is the antiferromagnet's resonance frequency, $H_s = H_e + \frac{1}{2}H_A$, γ is the magneto-mechanical ratio: $\gamma = eg/2m$, g is the gyromagnetic Lande factor and

$$\omega_L = \omega_T\sqrt{1 + \frac{M_s}{H_s}} \tag{7.5}$$

is the frequency of the long-wavelength longitudinal optical magnons.

The dielectric permittivity tensor of the antiferromagnet has the form

$$\varepsilon = \begin{pmatrix} \varepsilon_\perp & 0 & 0 \\ 0 & \varepsilon_\perp & 0 \\ 0 & 0 & \varepsilon_\parallel \end{pmatrix} \tag{7.6}$$

where $\varepsilon_\parallel$ and $\varepsilon_\perp$ are the frequency-independent dielectric constants in the directions parallel and perpendicular to the optical axis, respectively.

Using Eqs. (7.1), (7.2) and (7.6) for fields $\boldsymbol{E}$, $\boldsymbol{H} \sim \exp[i(k_\parallel x + k_z z - \omega t)]$, we obtain from the Maxwell equations two solutions for each medium: a TE-type wave with field components E_y, H_x, H_z and a TM-type wave with components H_y, E_x, E_z. The normal component of the wave vector in the semiconductor layer is the same for both waves:

$$k_z = i\sqrt{k_\parallel^2 - \frac{\omega^2}{c^2}\varepsilon(\omega)} \tag{7.7}$$

In the region occupied by the antiferromagnet, we obtain for TE and TM waves, respectively,

$$k_{zo} = -i\sqrt{\left(k_\parallel^2 - \frac{\omega^2}{c^2}\varepsilon_\perp\right)\mu_\perp(\omega)} \tag{7.8}$$

and

$$k_{ze} = -\mathrm{i}\sqrt{\left[k_{\|}^2 - \frac{\omega^2}{c^2}\varepsilon_{\|}\mu_{\perp}(\omega)\right]\frac{\varepsilon_{\perp}}{\varepsilon_{\|}}} \tag{7.9}$$

A necessary condition for the existence of nonradiative interface waves is that all three expressions under the square roots in Eqs. (7.7)–(7.9), and also $k_{\|}^2$, are real and positive.

From the boundary conditions at the surface of separation ($z = 0$) we obtain the dispersion relations for TM- and TE-type waves:

$$k_z = \frac{\varepsilon(\omega)}{\varepsilon_{\perp}} k_{ze} \tag{7.10}$$

and

$$k_z = \frac{k_{zo}}{\mu_{\perp}(\omega)} \tag{7.11}$$

respectively. Using Eqs. (7.7)–(7.10), we conclude that TM-type interface waves exist only if the conditions

$$\varepsilon(\omega) < 0 \qquad \frac{c^2 k_{\|}^2}{\omega^2} > \varepsilon_{\|}\mu_{\perp}(\omega) \tag{7.12}$$

are fulfilled, and they are described by the dispersion law

$$\frac{c^2 k_{\|}^2}{\omega^2} = \frac{\varepsilon_{\|}\varepsilon(\omega)[\varepsilon_{\perp} - \varepsilon(\omega)\mu_{\perp}(\omega)]}{\varepsilon_{\|}\varepsilon_{\perp} - \varepsilon^2(\omega)} \tag{7.13}$$

Similarly, we obtain the dispersion law for TE-type interface polaritons:

$$\frac{c^2 k_{\|}^2}{\omega^2} = \frac{\varepsilon(\omega)\mu_{\perp}(\omega) - \varepsilon_{\perp}}{\mu_{\perp}(\omega) - 1} \tag{7.14}$$

which exist under the conditions

$$\mu_{\perp}(\omega) < 0 \qquad \frac{\varepsilon_{\perp}}{\mu_{\perp}(\omega)} < \varepsilon(\omega) < \frac{c^2 k_{\|}^2}{\omega^2} < \varepsilon_{\perp} \tag{7.15}$$

In the case when the magnetic anisotropy axis is parallel to the interface, e.g. $\mathbf{C}\|oy$, for the permeability and permittivity tensors we have

$$\mu = \begin{pmatrix} \mu_\perp(\omega) & 0 & 0 \\ 0 & 1 & 0 \\ 0 & 0 & \mu_\perp(\omega) \end{pmatrix} \qquad \varepsilon = \begin{pmatrix} \varepsilon_\perp & 0 & 0 \\ 0 & \varepsilon_{||} & 0 \\ 0 & 0 & \varepsilon_\perp \end{pmatrix} \tag{7.16}$$

Using these tensors, we can easily verify that, for an arbitrary orientation of the two-dimensional wave vector $\boldsymbol{k}_{||}$ relative to the **C**-axis, the interface wave is, in general, a linear combination of an ordinary wave and an extraordinary wave. These two waves split only if the direction of propagation is either parallel or perpendicular to the **C**-axis. In the case $\boldsymbol{k}_{||}||\mathbf{C}$, there are no TE-type interface polaritons; the TM-type polaritons are characterized by a dispersion law and necessary conditions for their existence which differ from Eqs. (7.13) and (7.12) only by the substitution of $\varepsilon_{||}$ for $\varepsilon_\perp$ and *vice versa*. In the case $\boldsymbol{k}_{||}\perp\mathbf{C}$, there are both TE- and TM-type interface polaritons and their dispersion laws and conditions for their existence are, respectively,

$$\frac{c^2k_{||}^2}{\omega^2} = \frac{\mu_\perp(\omega)\left[\varepsilon(\omega)\mu_\perp(\omega) - \varepsilon_{||}\right]}{\mu_\perp^2(\omega) - 1}, \qquad \mu_\perp(\omega) < 0, \quad \frac{c^2k_{||}^2}{\omega^2} > \varepsilon(\omega) \tag{7.17}$$

$$\frac{c^2k_{||}^2}{\omega^2} = \frac{\varepsilon_\perp\varepsilon(\omega)}{\varepsilon_\perp + \varepsilon(\omega)}, \qquad \varepsilon(\omega) < 0, \quad |\varepsilon(\omega)| > \varepsilon_\perp \tag{7.18}$$

The wave described by Eq. (7.18) does not interact with the optical magnons and is coupled only to the surface plasmons. The interface magnon–plasmon polaritons correspond to the waves with dispersion laws (7.13), (7.14) and (7.17). We shall further consider only such waves.

7.2
Properties of TM-type Interface Magnon–Plasmon Polaritons

The propagation of TM-type interface polaritons described by Eq. (7.13) is entirely due to the presence of the free carriers because one of the conditions (7.12), namely $\varepsilon(\omega) < 0$ would be violated in the absence of the free carriers. Substituting Eqs. (7.1) and (7.3) into Eq. (7.13), we obtain a biquartic equation for the frequency ω. Analyzing this equation, we easily arrive at the conclusion that only one of its solutions satisfies the conditions (7.12) if the free carrier density N in the semiconductor is such that the plasma frequency $\omega_p < \omega_T$. This means that only a single TE mode is possible. The dispersion curve of this mode in the $(k_{||}, \omega)$-plane begins at the point $k_{||} = \omega = 0$ and increases monotonically with increasing $k_{||}$ and, for $k_{||} >> \omega/c$ tends asymptotically to the line $\omega = \varpi_p$, where

$$\varpi_p = \frac{\omega_p}{\sqrt{1 + \left(\varepsilon_{||}\varepsilon_\perp\right)^{1/2}\varepsilon_\infty^{-1}}} \tag{7.19}$$

For densities N such that $\omega_p > \omega_T$, a second TM mode appears whose dispersion curve begins at the point with coordinates $k_{\|} = 0,\ \omega = \omega_+$ (see later); it decreases or increases monotonically with increasing $k_{\|}$ and tends asymptotically to the line $\omega = \max\{\omega_T, \varpi_p\}$ as $k_{\|} \to \infty$. The limiting frequency of this mode for $k_{\|} = 0$ is determined by the solution ω_- (provided that $\varepsilon_\perp \neq \varepsilon_\infty$) of the biquadratic equation $\varepsilon(\omega)\mu_\perp(\omega) = \varepsilon_\perp$:

$$\omega_\pm^2 = \frac{b \pm \sqrt{b^2 - 4\omega_p^2\omega_L^2\left(1 - \varepsilon_\perp\varepsilon_\infty^{-1}\right)}}{2\left(1 - \varepsilon_\perp\varepsilon_\infty^{-1}\right)}, \qquad b \equiv \omega_L^2 + \omega_p^2 - \frac{\varepsilon_\perp}{\varepsilon_\infty}\omega_T^2 \tag{7.20}$$

In the case $\varepsilon_\perp = \varepsilon_\infty$, the frequency ω_- is replaced by

$$\omega_0 = \frac{\omega_p\omega_L}{\sqrt{\omega_p^2 + \omega_L^2 - \omega_T^2}} \tag{7.21}$$

Both ω_- and ω_0 increase monotonically with increasing N, remaining always lower than ω_L.

Thus, the number of TM-type interface modes depends on the carrier density. Such a dependence was first predicted in Ref. [61] for the surface polaritons propagating along the interface between an anisotropic semiconductor and a vacuum (see Chapter 5).

Figure 7.1 shows the dispersion curves of TM modes described by Eq. (7.13), for various carrier densities. For simplicity it has been assumed that $\varepsilon_\perp = \varepsilon_{\|} = \varepsilon_\infty$, so that $\varpi_p = \omega_p/\sqrt{2}$. It can be seen from Fig. 7.1 that the high-frequency limit of the fundamental (first) mode increases with increasing N and reaches eventually a fixed mode which appears only for $\varpi_p \geq \omega_T$. The dispersion curve of the upper mode which appears only for $\omega_p > \omega_T$ also shifts toward higher frequencies with increasing N and, in the situation shown in Fig. 7.1b and c, we have $d\omega/dk_{\|} < 0$, that is, the group and phase velocities of the waves are oriented in opposite directions. However, such anomalous dispersion is obtained only for $N < N_{cr}$ where the critical value of the carrier density is given by

$$\omega_p^2(N_{cr}) = \omega_T^2 + \omega_L^2 \tag{7.22}$$

In the case $N > N_{cr}$ the dispersion of the upper mode has the usual form: $d\omega/dk_{\|} > 0$ (see Fig. 7.1d). It is of interest that the gap in the spectrum of the interface polaritons in the region of anomalous dispersion first narrows with increasing N, vanishes for $\varpi_p = \omega_T$, and then appears again and widens, remaining always narrower than the gap for bulk optical magnons $\omega_L - \omega_T$.

It follows that not only the existence but also the main properties of the interface polaritons, that is, the number of the modes, the position and width of their frequency range, the type of their dispersion, the width of the gap and so forth, depend strongly on the density of the free carriers. This result can be used to

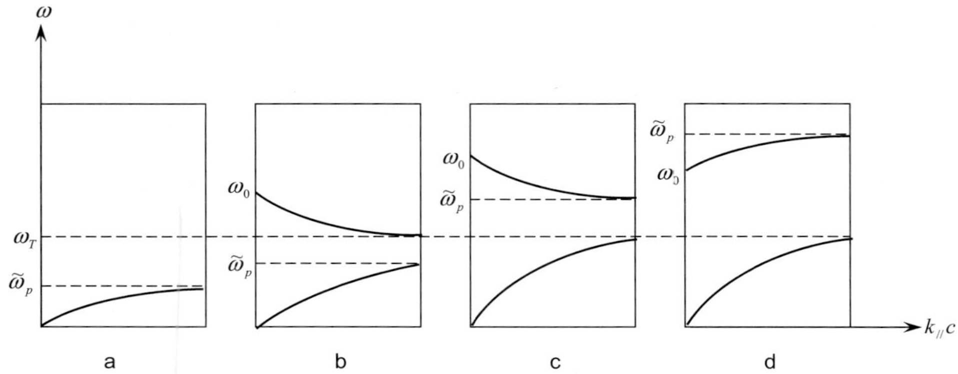

Fig. 7.1 Dispersion curves for TM-type interface polaritons, for various densities of the free charge carriers:
a. $\omega_p < \omega_T$, b. $\omega_T < \omega_p < \sqrt{2}\omega_T$,
c. $\omega_T\sqrt{2} < \omega_p < \sqrt{\omega_T^2 + \omega_L^2}$, d. $\omega_p > \sqrt{\omega_T^2 + \omega_L^2}$.

construct delay lines and other devices with specified properties of the surface waves based on two-layer structures consisting of an antiferromagnetic insulator and a semiconductor.

7.3
Effect of Free Carriers on the Properties of TE-type Interface Polaritons

When the easy axis of the antiferromagnet is perpendicular to the interface with the semiconductor, the dispersion properties of TE-type interface polaritons are described by Eq. (7.14), which is biquadratic in ω. However, only one of the solutions of this equation satisfies the condition (7.15), that is, there is only a single TE mode whose frequency is given by

$$\omega^2 = \frac{b - \sqrt{b^2 - 4(1 - \varepsilon_\perp \varepsilon_\infty^{-1})\left[\omega_L^2\omega_p^2 + c^2k_{||}^2(\omega_L^2 - \omega_T^2)\varepsilon_\infty^{-1}\right]}}{2(1 - \varepsilon_\perp \varepsilon_\infty^{-1})} \tag{7.23}$$

if $\varepsilon_\perp \neq \varepsilon_\infty$, and

$$\omega^2 = \frac{\omega_L^2\omega_p^2 + c^2k_{||}^2(\omega_L^2 - \omega_T^2)}{\omega_p^2 + \omega_L^2 - \omega_T^2} \tag{7.24}$$

for $\varepsilon_\perp = \varepsilon_\infty$; b is defined in Eq. (7.20). The frequency of the mode increases monotonically with increasing $k_{||}$; the latter varies within a bounded interval

$k_1 < k_{||} < k_2$. The frequency range of the wave is also bounded: $\omega_1 < \omega < \omega_2$. The boundaries of these regions and, therefore, also their widths depend strongly on the parameters of both media in contact. We list in Table 7.1 the values of these limits for various values of the dimensionless parameters $\varepsilon_\perp/\varepsilon_\infty$, ω_p/ω_T and ν, where we use the following notation:

$$\nu = \sqrt{\left(1 - \frac{\varepsilon_\perp}{\varepsilon_\infty}\right)\left(1 + \frac{M_s}{H_s}\right)} \qquad \nu_1 = \frac{\omega_p}{c}\sqrt{\frac{\varepsilon_\perp \varepsilon_\infty}{\varepsilon_\infty - \varepsilon_\perp}} \tag{7.25}$$

$$\nu_2 = \frac{\omega_T}{c}\sqrt{\varepsilon_\perp\left(1 + \frac{M_s}{H_s}\right)} \qquad \nu_3 = \frac{\omega_T}{c}\sqrt{\varepsilon_\infty\left(1 - \frac{\omega_p^2}{\omega_T^2}\right)} \tag{7.26}$$

Table 7.1 Limiting values of the frequency ω (s^{-1}) and the wave number $k_{||}$ (m^{-1}) of TE-type interface modes for various values of the parameters $\varepsilon_\perp/\varepsilon_\infty$, ω_p/ω_T and ν. Expressions for $\omega_\pm$ are given by Eq. (7.20).

	$\varepsilon_\perp < \varepsilon_\infty$						$\varepsilon_\perp > \varepsilon_\infty$	
	$\nu < 1$			$\nu > 1$				
	$\sqrt{1 - \frac{\varepsilon_\perp}{\varepsilon_\infty}} < \frac{\omega_p}{\omega_T} < \nu$	$\nu \le \frac{\omega_p}{\omega_T} < 1$	$\frac{\omega_p}{\omega_T} \ge 1$	$\sqrt{1 - \frac{\varepsilon_\perp}{\varepsilon_\infty}} < \frac{\omega_p}{\omega_T} \le 1$	$1 < \frac{\omega_p}{\omega_T} < \nu$	$\frac{\omega_p}{\omega_T} \ge \nu$	$\frac{\omega_p}{\omega_T} \le 1$	$\frac{\omega_p}{\omega_T} > 1$
ω_1	ω_T	ω_T	ω_+	ω_T	ω_+	ω_+	ω_T	ω_-
ω_2	$\omega_p(1 - \varepsilon_\perp\varepsilon_\infty^{-1})^{-1/2}$	ω_L	ω_L	$\omega_p/\sqrt{1 - \varepsilon_\perp\varepsilon_\infty^{-1}}$	$\omega_p/\sqrt{1 - \varepsilon_\perp\varepsilon_\infty^{-1}}$	ω_L	ω_L	ω_L
k_1	ν_3	ν_3	0	ν_3	0	0	ν_3	0
k_2	ν_1	ν_2	ν_2	ν_1	ν_1	ν_2	ν_2	ν_2

It is of interest to investigate how the region of existence of the mode under consideration varies as a function of the free carrier density N. First of all, we have to note that this mode in the structures with $\varepsilon_\perp < \varepsilon_\infty$ exists even in the absence of the free carriers, in contrast to TM modes, provided the values of ω and $k_{||}$ lie in the regions

$$\omega_T < \omega < \omega_L \qquad \frac{\omega_T}{c}\sqrt{\varepsilon_\infty} < k_{||} < \nu_2 \tag{7.27}$$

In the presence of the free carriers, the lower limit of the wave number k_1 decreases with increasing N and infinite wavelengths become possible for

$\omega_p > \omega_T \quad (k_\| \to 0)$. Further increase in N gives rise to an increase in ω_1 but the width of the frequency range decreases and may become quite small. For $\omega_p > \omega_T\sqrt{1-\varepsilon_\perp\varepsilon_\infty^{-1}}$, the wave may also be propagated in the structures with $\varepsilon_\perp < \varepsilon_\infty$, for example in the CdMnTe/CdTe two-layer structure, where $\varepsilon_\perp = 5$ and $\varepsilon_\infty = 10$. The limits of the existence region then depend not only on N but also on the parameter ν (see Table 7.1).

Let us now briefly discuss the effect of carriers on the properties of TE-type modes described by Eq. (7.17). It can be seen that there is again only one mode of this type and its dispersion curve for $\omega_p < \omega_T$ begins at the point $\omega = \omega_T$, $k_\| = \nu_3$, increases monotonically with increasing $k_\|$ and approaches asymptotically the line $\omega = \omega_3$ as $k_\| \to \infty$ where

$$\omega_3 = \omega_T\sqrt{1+\frac{\omega_T^2}{2(\omega_L^2-\omega_T^2)}} \tag{7.28}$$

We note that the width of the interval of admissible frequencies for $\omega_p < \omega_T$ has no dependence on N and is only determined by characteristic frequencies of the optical magnons, but k_1 decreases with increasing N and vanishes for $\omega_p = \omega_T$. As N increases further so that $\omega_T < \omega_p < \omega_3$ the region of existence narrows since $\omega_1 = \omega_+$ increases. For $\omega_p > \omega_3$ the dispersion of the wave becomes anomalous, that is, the frequency decreases as a function of $k_\|$ from $\omega = \omega_+$ corresponding to $k_\| = 0$ to $\omega = \omega_3$ as $k_\| \to \infty$ while the width of the existence region increases with increasing N but always remains smaller than $\Delta\omega = \omega_L - \omega_3$. It follows that there is again a critical value N_{cr} such that the group velocity of the wave for $N < N_{cr}$ satisfies $d\omega/dk_\| > 0$ whereas $d\omega/dk_\| < 0$ for $N > N_{cr}$. This value is equal to half the critical density for the high-frequency TM mode [Eq. (7.22)] and is given by

$$N_{cr} = \frac{m * \varepsilon_\infty \omega_T^2}{e^2}\left(1+\frac{M_s}{2H_s}\right) \tag{7.29}$$

Using the data for FeF_2 in our estimates: $\omega_T = 1.58 \times 10^{12} s^{-1}$, $M_s/H_s = 0.6$, and assuming that $m^* = 0.1m$, $\varepsilon_\infty = 10$ we obtain $N_{cr} \approx 5 \times 10^{14} cm^{-3}$.

7.4
Reflection Coefficient in the Method of Frustrated Total Internal Reflection

The most frequently used method of exciting surface waves is the method of frustrated total internal reflection (FTIR), which reduces to measurements of the frequency dependence of the reflection coefficient for the electromagnetic waves incident on the base of a FTIR prism. In Otto's approach [45], a thin layer of air separates the bulk surface-active medium from the prism whereas, in Kretschmann's experiments [74], a thin layer of surface-active medium was deposited directly on the base of the FTIR prism. Only calculations of the reflection coefficient R for an interface between a solid and a vacuum and only for TM waves were available in Refs. [48, 50, 72]. A generalization of such calculations to the excitation of TM- and TE-type interface polaritons at the interface between an antiferromagnetic insulator and a semiconductor has been made in Ref. [68]. In the case of TM-type interface polaritons described by Eq. (7.13), the surface-active medium is the semiconductor layer. In Otto's method, a thin layer of an antiferromagnet of thickness d_2 is between the semiconductor and the prism (an isotropic medium with permittivity $\varepsilon_p > \varepsilon$).

Let us assume that a p-polarized electromagnetic wave is incident on the side of the prism and penetrates across the thin gap and reaches the interface between the antiferromagnetic insulator and the semiconductor. After fairly simple but lengthy calculations using standard boundary conditions for the fields at both interfaces, we obtain the following expression for the reflection coefficient:

$$R_p^0 = 1 - \frac{4\varepsilon_{\|}\varepsilon_{\perp}\sqrt{\varepsilon_p}\cos\theta\exp(-2\omega\kappa d_2/c)}{\varepsilon_p^2\sin^2\theta + \varepsilon_{\|}\varepsilon_{\perp}\cos^2\theta - \varepsilon_{\|}\varepsilon_p\mu_{\perp}(\omega)}\delta\left[\frac{\varepsilon_{\perp}}{\kappa} + \frac{\varepsilon(\omega)}{\kappa_1}\right] \tag{7.30}$$

where

$$\kappa = \sqrt{\varepsilon_{\perp}\varepsilon_{\|}^{-1}\left[\varepsilon_p\sin^2\theta - \varepsilon_{\|}\mu_{\perp}(\omega)\right]} \qquad \kappa_1 = \sqrt{\varepsilon_p\sin^2\theta - \varepsilon(\omega)} \tag{7.31}$$

and θ is the angle of incidence. The delta function appears in Eq. (7.30) because we neglect dissipation. Note that $R_p^O < 1$ if

$$\frac{\varepsilon_{\perp}}{\kappa} + \frac{\varepsilon(\omega)}{\kappa_1} = 0 \tag{7.32}$$

Noting also that

$$\varepsilon_p\sin^2\theta = \frac{c^2k_{\|}^2}{\omega^2} \tag{7.31a}$$

it can be easily seen that the condition (7.32) is identical to the dispersion relation (7.13). It means that a sharp reduction in the reflection intensity corresponds to excitation of TM-type interface polaritons.

In the case of using Kretschmann's method for excitation of the waves, we need to determine the reflection coefficient in the structure FTIR prism/semiconductor of thickness d_1/antiferromagnet. The corresponding formula has the form

$$R_p^K = 1 - \frac{4\varepsilon^2(\omega)\cos\theta\exp(-2\omega\kappa_1 d_1/c)}{\varepsilon_p^{1/2}\left[\varepsilon_p^2\sin^2\theta + \varepsilon^2(\omega)\cos^2\theta - \varepsilon(\omega)\varepsilon_p\right]}\delta\left[\frac{\varepsilon_{\perp}}{\kappa} + \frac{\varepsilon(\omega)}{\kappa_1}\right] \tag{7.33}$$

In addition to the inequalities (7.12), which follow from the conditions that κ and κ_1 are real and positive, we require in both cases that the condition of transparency of the prism ($c^2k_{\|}^2/\omega^2 < \varepsilon_p$) as well as the relation (7.32) is satisfied.

In the case of TE-type interface polaritons described by Eq. (7.14), the surface-active medium is the layer of the antiferromagnet. Assuming that the semiconductor film is again located between the prism and the antiferromagnetic layer, we obtain in Otto's method for an s-polarized incident wave the following expression for the reflection coefficient:

$$R_s^O = 1 - \frac{4\varepsilon_p^{1/2}\kappa_1^2\cos\theta\exp(-2\omega\kappa_1 d_1/c)}{\varepsilon_p - \varepsilon(\omega)}\delta\left[\kappa_1 + \frac{\kappa_2}{\mu_\perp(\omega)}\right] \tag{7.34}$$

where

$$\kappa_2 = \sqrt{\left(\varepsilon_p\sin^2\theta - \varepsilon_\perp\right)\mu_\perp(\omega)} \tag{7.35}$$

In Kretschmann's method, positioning the antiferromagnetic layer between the prism and the semiconductor layer, we obtain

$$R_s^K = 1 - \frac{4\varepsilon_p^{1/2}\kappa_1^2\cos\theta\exp(-2\omega\kappa_2 d2/c)}{1 + \varepsilon_p\mu_\perp(\omega)\left(\varepsilon_p\sin^2\theta - \varepsilon_\perp\right)^{-1}\cos^2\theta}\delta\left[\kappa_1 + \frac{\kappa_2}{\mu_\perp(\omega)}\right] \tag{7.36}$$

Both R_s^O and R_s^K differ from unity if the frequency and angle of incidence satisfy the equation

$$\kappa_1 + \frac{\kappa_2}{\mu_\perp(\omega)} = 0 \tag{7.37}$$

which is identical with the dispersion equation (7.14) for TE-type interface polaritons.

The expressions (7.34) and (7.36) also remain valid for TE-type interface polaritons described by Eq. (7.17) provided that we replace κ_2 by κ_2', where

$$\kappa_2' = \sqrt{\varepsilon_p\sin^2\theta - \varepsilon_{\|}\mu_\perp(\omega)} \tag{7.38}$$

Note that in all the considered cases, the position of the minimum of the reflection coefficient $R(\omega)$ is independent of the thickness d of the gap. It follows that d should be chosen in experiments in such a way that this condition is not violated.

7.5 Complete Transmission of Electromagnetic Waves by a Two-Layer Structure

Let us consider now the effect of complete transmission (without reflection) of electromagnetic waves by a two-layer structure consisting of an antiferromagnetic insulator and a semiconductor under resonant excitation of the interface magnon–plasmon polaritons [68]. We will assume that the semiconductor layer occupies the space region $0<z<d_1$, the region $-d_2<z<0$ is occupied by the antiferromagnetic layer and the regions $z>d_1$, $z<-d_2$ are filled with an isotropic medium whose permittivity is ε_p (prisms).

Let us assume that a plane-polarized wave is incident obliquely on the two-layer structure from the region $z<-d_2$ (or $z>d_1$) and the fields of the wave are $\boldsymbol{E}$, $\boldsymbol{H}$ ~$\exp[i(k_{\|}x+k_z z-\omega t)]$. We will consider separately the cases of s- and p-polarized waves.

a) Assume that the incident wave is s-polarized: $\boldsymbol{E}=(0, E_y, 0)$, $\boldsymbol{H}=(H_x, 0, H_z)$. In the case when the orientation of the magnetic anisotropy axis satisfies $\mathbf{c}\|oz$, we obtain the following expression for the reflection coefficient:

$$R_s=\frac{\left[a_s(\kappa_1\kappa_2+\mu_\perp\kappa_0^2)(\kappa_2+\mu_\perp\kappa_1)+b_s(\kappa_1\kappa_2-\mu_\perp\kappa_0^2)(\kappa_2-\mu_\perp\kappa_1)\right]^2+S_-^2(\kappa_2^2-\mu_\perp^2\kappa_1^2)}{\left[a_s(\kappa_1\kappa_2-\mu_\perp\kappa_0^2)(\kappa_2+\mu_\perp\kappa_1)+b_s(\kappa_1\kappa_2+\mu_\perp\kappa_0^2)(\kappa_2-\mu_\perp\kappa_1)\right]^2+\left[S_-(\kappa_2^2+\mu_\perp^2\kappa_1^2)+2S_+\mu_\perp\kappa_1\kappa_2\right]^2} \tag{7.39}$$

where

$$a_s=g_1g_2-1 \qquad b_s=g_2-g_1 \qquad g_i=\exp(-2\omega\kappa_i d_i/c) \quad i=1,2; \tag{7.39a}$$

$$S_\pm=\kappa_0(1\pm g_1)(1\pm g_2) \qquad \kappa_0=\sqrt{\varepsilon_p-c^2k_{\|}^2/\omega^2} \tag{7.39b}$$

It follows from Eq. (7.39) that the wave can be transmitted completely by the two-layer structure provided that the conditions

$$\kappa_2+\mu_\perp\kappa_1=0 \quad \kappa_1 d_1-\kappa_2 d_2=0 \tag{7.40}$$

are satisfied simultaneously; we then have $R_s=0$

The first equation of the system (7.40) is identical with the dispersion law (7.14). The condition $\mu_\perp(\omega)<0$, which follows from this equation for $\kappa_1>0$ and $\kappa_2>0$ determines the range of frequencies $\omega_T<\omega<\omega_L$ where the incident wave would be totally reflected from the antiferromagnet in the absence of the semiconductor layer. Such a gap in the spectrum of an antiferromagnet was observed, for example, in CoF_3 crystals in the far-infrared region [75]. It is the excitation of the TE-type surface polaritons at the interface between the layers that ensures the transmission of the s-polarized wave across the nontransparent region.

The second equation of the system (7.40) ensures that the wave is transmitted completely, that is, the incident and transmitted energy fluxes are equal.

It should be noted that, because of the wave interference, the amplitude of the field of the wave decreases with increasing distance from the interface $z = 0$ much more slowly than the usual exponential law:

$$E_{yi}(z) = E_0 \left[\cosh^2 \varphi_i + \left(\frac{\kappa_0}{\kappa_i} \right)^2 \sinh^2 \varphi_i \right]^{1/2} \tag{7.41}$$

where $\varphi_i = \kappa_i z \omega / c \quad i = 1, 2$, and E_0 is the amplitude of the incident wave.

Examining the compatibility of Eqs. (7.40), we obtain $\mu_\perp(\omega) = -d_1/d_2$ and, using Eqs. (7.3) and (7.14), we find that there are fixed values of the frequency Ω_0 and the angle of incidence θ_0 corresponding to the complete transmission (perfect transparency):

$$\Omega_0^2 = \frac{d_1 \omega_T^2 + d_2 \omega_L^2}{d_1 + d_2} \tag{7.42}$$

$$\sin^2 \theta_o = \frac{\varepsilon_s}{\varepsilon_p} \left(1 - \frac{\omega_p^2}{a_0 \Omega_0^2} \right) \tag{7.43}$$

where

$$\varepsilon_s = \frac{d_1 \varepsilon_\infty + d_2 \varepsilon_\perp}{d_1 + d_2} \qquad a_0 = 1 + \frac{d_2 \varepsilon_\perp}{d_1 \varepsilon_\infty} \tag{7.43a}$$

It follows that a complete transmission of an s-polarized electromagnetic wave of frequency Ω_0 by the two-layer structure is accompanied by the excitation of TE-type interface polaritons described by Eq. (7.14) with resonance frequency $\omega = \Omega_0$ and wave number given by

$$k_{||} c = \sqrt{\varepsilon_s \left(\Omega_0^2 - a_0^{-1} \omega_p^2 \right)} \tag{7.44}$$

Note that the presence of the free carriers in the semiconductor layer does not affect the value of Ω_0 but modifies considerably the angle θ_0, which depends on the carrier density. Note also that the perfect transparency of a nonconducting two-layer structure can occur for

$$\varepsilon_p > \varepsilon_s \quad \varepsilon_\perp > \varepsilon_\infty \tag{7.45}$$

whereas these conditions are not necessary in the case under study. Moreover, it is found that the perfect transparency of a two-layer structure is only possible in a limited range of carrier densities

$$N_{min} < N < N_0, \tag{7.46}$$

where N_0 can be obtained from the condition

$$\omega_p^2(N_0) = a_0\Omega_0^2 \tag{7.47}$$

and N_{min} may assume three different values: 0, N_1 or N_2 depending on the parameters $\varepsilon_p, \varepsilon_\perp, \varepsilon_s$ and ε_∞ ; here

$$N_1 = N_0\left(1 - \frac{\varepsilon_p}{\varepsilon_s}\right) \qquad N_2 = \frac{N_0}{a_0}\left(1 - \frac{\varepsilon_\perp}{\varepsilon_\infty}\right) \tag{7.48}$$

and we have $N_{min} = N_1$ if $\varepsilon_p < \min\{\varepsilon_s, \varepsilon_\perp\}$ and $N_{min} = N_2 < N_1$ if $\varepsilon_\perp < \min\{\varepsilon_p, \varepsilon_\infty\}$ In all other cases, $N_{min} = 0$ and the carrier density is only bounded from above.

We note that the angle θ_0 and, therefore, also the wave number $k_{||}$ decrease monotonically with increasing N in the region (7.46) and vanish for $N = N_0$. The maximum values of these quantities for $N = N_{min}$ are given by

$$\sin^2\theta_0^{max} = \frac{\gamma}{\varepsilon_p} \qquad k_{||}^{max} = \frac{\Omega_0}{c}\sqrt{\gamma} \tag{7.49}$$

where $\gamma = \varepsilon_s, \varepsilon_p$ or $\varepsilon_\perp$ corresponding, respectively, to $N_{min} = 0$, N_1 or N_2.

Let us now consider briefly the case when the c-axis is parallel to the interface. For $c||k_{||}$, the zero-reflection transmission of a s-polarized wave cannot occur. For $c \perp k_{||}$, Eqs. (7.39)–(7.41) remain valid provided that we replace κ_2 by κ_2' [see Eq. (7.32)]. The condition $R_s = 0$ then yields a dispersion law for TE-type interface polaritons described by Eq. (7.17). The complete transmission of an electromagnetic wave occurs again at the same frequency Ω_0 [Eq. (7.42)] but for another angle of incidence, which differs from θ_0 by the substitution of a_1 for a_0 and ε_{s1} for ε_s in Eq. (7.43), where

$$a_1 = 1 + \frac{d_2\varepsilon_{||}}{d_1\varepsilon_\perp} \qquad \varepsilon_{s1} = \frac{\varepsilon_\infty a_1 d_1^2}{d_1^2 - d_2^2} \tag{7.50}$$

In contrast to the case $c||oz$, a complete transmission of the incident wave under resonant excitation of TE-type interface polaritons described by Eq. (7.17) can only occur for $d_1 \neq d_2$ and the admissible values of the carrier density N for $d_1 > d_2$ are

$$N_1 < N < N_0 \tag{7.51}$$

and, in the case $d_1 < d_2$, higher densities are needed:

$$N_0 < N < N_1' \tag{7.52}$$

Here,

$$\omega_p^2(N_0) = a_1\Omega_0^2 \qquad N_1 = N_0\left(1 - \frac{\varepsilon_p}{\varepsilon_{s1}}\right) \qquad N_1' = N_0\left(1 + \frac{\varepsilon_p}{\varepsilon_{s1}}\right) \tag{7.53}$$

To have some idea about the numerical values of the frequency Ω_0 and of the critical densities, we will again use the parameters for FeF_2 (see Ref. [73]). Bearing in mind that $\varepsilon_\perp = 5$ and $\varepsilon_\| = 2$ for this material and setting $d_1/d_2 = 0.5$ and $\varepsilon_p = 1$ we obtain $\Omega_0 \approx 1.8 \times 10^{12} s^{-1}$ and $N_0 = 1.9$, $N_1 = 2.2$ in units of 10^{15} cm^{-3}

b) Let us assume now that a p-polarized wave with the vector $\boldsymbol{E}$ lying in the plane of incidence xz is incident on the two-layer structure. Using standard continuity conditions for the fields at the interfaces $z = d_1$, $z = 0$ and $z = -d_2$ in the case $\mathbf{c}||oz$, we obtain the following expression for the reflection coefficient:

$$R_p = \frac{\left(a_p f_{1+} f_{2+} + b_p f_{1-} f_{2-}\right)^2 + \varepsilon_p^2 f_{1+} f_{1-} P_-^2}{\left(a_p f_{1+} f_{2-} + b_p f_{1-} f_{2+}\right)^2 + \varepsilon_p^2 \left[P_-\left(\varepsilon_\perp^2 \kappa_1^2 + \varepsilon^2\kappa^2\right) + 2P_+ \varepsilon\varepsilon_\perp \kappa\kappa_1\right]^2} \tag{7.54}$$

where

$$a_p = g_1 g_3 - 1 \qquad b_p = g_3 - g_1 \qquad g_3 = \exp(-2\omega\kappa d_2/c) \tag{7.54a}$$

$$P_\pm = \kappa_0(1 \pm g_1)(1 \pm g_3) \qquad f_{1\pm} = \varepsilon_\perp \kappa_1 \pm \kappa\varepsilon(\omega) \qquad f_{2\pm} = \varepsilon_p^2 \kappa\kappa_1 \pm \kappa_0^2 \varepsilon_\perp \varepsilon(\omega) \tag{7.54b}$$

Here κ and κ_1 are given by Eq. (7.31), and g_1 and κ_0 by Eqs. (7.39a) and (7.39b).

It follows from Eq. (7.54) that a complete transmission of a p-polarized electromagnetic wave ($R_p = 0$) occurs if the conditions

$$\varepsilon_\perp \kappa_1 + \kappa\varepsilon(\omega) = 0 \qquad d_1\kappa_1 - d_2\kappa = 0 \tag{7.55}$$

are satisfied. The first equation in Eqs. (7.55) can be transformed to the dispersion equation for TM-type interface polaritons given by Eq. (7.13). The requirement of compatibility of the conditions (7.55) yields $\varepsilon(\omega) = -\varepsilon_\perp d_2/d_1$ and, using this equality together with Eqs. (7.1) and (7.13), we obtain the following expressions for the perfect transparency frequency Ω_1 and for the angle of incidence θ_1:

$$\Omega_1 = \frac{\omega_p}{\sqrt{1 + \frac{\varepsilon_\perp d_2}{\varepsilon_\infty d_1}}} \tag{7.56}$$

$$\sin^2 \theta_1 = \frac{\varepsilon_d \left(\Omega_1^2 - \Omega_0^2\right)}{\varepsilon_p \left(\Omega_1^2 - \omega_T^2\right)} \tag{7.57}$$

where

$$\varepsilon_d = \frac{\varepsilon_{||}\varepsilon_\perp d_2(d_1 + d_2)}{\varepsilon_\perp d_2^2 - \varepsilon_{||} d_1^2} \tag{7.57a}$$

The perfect transparency of the two-layer structure for a p-polarized wave is accompanied by the excitation of TM-type interface polaritons (7.13) of frequency Ω_1 and with a two-dimensional wave number $k_{||} = \Omega_1 \sqrt{\varepsilon_p} \sin \theta_1 / c$. Setting, for example, $N = 10^{16}\,\text{cm}^{-3}$ and using again the same values of the characteristic parameters, we obtain $\Omega_1 = 3.86 \times 10^{12} \text{s}^{-1}$, $\theta_1 = 30°$ and $k_{||} = 64\,\text{cm}^{-1}$.

It can be seen that the complete transmission of a p-polarized wave by the two-layer structure is impossible in the absence of the free carriers. When carriers are present, the effect can take place provided that

$$\frac{d_2}{d_1} \neq \sqrt{\frac{\varepsilon_{||}}{\varepsilon_\perp}} \tag{7.58}$$

Moreover, the carrier density must be lie within well-defined intervals whose number and limits depend on the condition (7.58). In the case $d_2/d_1 < \sqrt{\varepsilon_{||}/\varepsilon_\perp}$ there is only one interval

$$N_+ < N < N_0 \tag{7.59}$$

while there are two regions of carrier density when the opposite inequality is satisfied:

$$N < N_-, \quad N > N_0 \tag{7.60}$$

provided that $\varepsilon_p > \varepsilon_d \Omega_0^2/\omega_T^2$, and only one region $N > N_0$ if $\varepsilon_p \leq \varepsilon_d \Omega_0^2/\omega_T^2$. The limiting values $N_\pm$ are determined by the following relations:

$$\omega_p^2(N_\pm) = a_0 \frac{\varepsilon_p \omega_T^2 \pm |\varepsilon_d| \Omega_0^2}{\varepsilon_p \pm |\varepsilon_d} \tag{7.61}$$

where a_0 is given in Eq. (7.43a).

As the carrier density varies within the regions (7.59) and (7.60), the frequency of perfect transparency Ω_1 varies in the intervals

$$\Omega_+ < \Omega_1 < \Omega_o, \qquad \Omega_1 < \Omega_-, \qquad \Omega_1 > \Omega_0 \tag{7.62}$$

respectively. It follows that complete transmission of a p-polarized wave by the structure is possible at both lower and higher frequencies than in the case of an s-polarized wave. Another special feature of p-polarized waves is that the zero-reflection regime cannot be realized for any choice of structure parameters in the frequency range

$$\Omega_- < \Omega < \Omega_+ \tag{7.63}$$

since this region lies within the gap of the spectrum of TM-type interface polaritons described by Eq. (7.13).

It should be noted that, if two modes of TM-type interface polaritons exist, the complete transmission of an electromagnetic wave is linked only to the resonant excitation of the upper (high-frequency) mode (see Fig. 7.1); the low-frequency TM mode is excited for $N < N_-$. In this case, there is no upper mode (Fig. 7.1a).

When the *c*-axis is parallel to the interface, Eqs. (7.54) and (7.55) remain valid if we replace κ by $\kappa' = \sqrt{c^2 k_{||}^2/\omega^2 - \varepsilon_\perp}$. In this case ($\mathbf{k}_{||} \perp \mathbf{c}$), the frequency and angle of incidence of the perfect transparency differ from the values of Ω_1 and θ_1 defined by Eqs. (7.56) and (7.57) by the substitution of $\varepsilon_\perp$ for $\varepsilon_{||}$ and *vice versa*. In the case $\mathbf{k}_{||} || \mathbf{c}$, the zero-reflection regime is realized at the same frequency Ω_1 but the angle of incidence is independent of the carrier density:

$$\sin^2\theta_2 = \frac{\varepsilon_\perp d_2}{\varepsilon_\mathrm{p}(d_2 - d_1)} \tag{7.64}$$

unlike the value of θ_1. It follows from Eq. (7.64) that this effect can take place if

$$\frac{d_1}{d_2} + \frac{\varepsilon_\perp}{\varepsilon_\mathrm{p}} < 1 \tag{7.65}$$

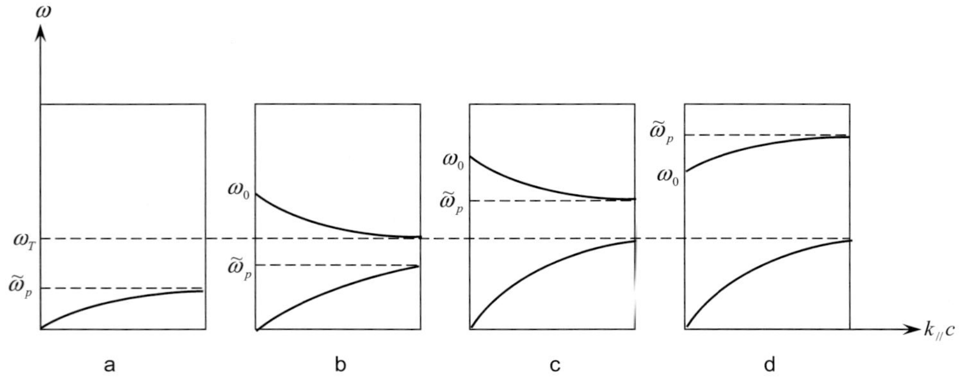

Fig. 7.1 Dispersion curves for TM-type interface polaritons, for various densities of the free charge carriers:
a. $\omega_p < \omega_T$, b. $\omega_T < \omega_p < \sqrt{2}\omega_T$, c. $\omega_T\sqrt{2} < \omega_p < \sqrt{\omega_T^2 + \omega_L^2}$, d. $\omega_p > \sqrt{\omega_T^2 + \omega_L^2}$.

that is, if the semiconductor layer is thinner than the antiferromagnetic layer with $\varepsilon_\perp < \varepsilon_p$. The complete transmission of a p-polarized wave is then accompanied by excitation of TM-type plasmon–polaritons described by Eq. (7.18).

We note that the effects studied in this section can be used in polarizers and filters for both s- and p-polarized electromagnetic waves with a specified frequency Ω_0 or Ω_1 for a suitable choice of d_1, d_2 and N. The relative width of the transmission band of such a filter is determined by the quantity $\exp(-2\omega d_1 \kappa_1/c)$ and can be quite small for the optimum choice of d_1.

Finally, we shall summarize our main conclusions in this chapter before going on to the study of more complicated structures.

1. The coupled magnon–plasmon interface polaritons of types TE and TM can travel along the interface between a semiconductor and a uniaxial antiferromagnet in the case when the plasmon frequency in the semiconductor becomes comparable to the optical magnon frequency in the antiferromagnet.
2. In contrast to TM-type waves whose existence depends upon the presence of free carriers, TE-type interface waves can exist even in the absence of carriers.
3. The number of modes, their dispersion and the width of the gap in the spectrum of TM-type interface polaritons depend strongly on the density of free carriers. If the density is such that the plasma frequency ω_p is lower than the antiferromagnetic resonance frequency ω_T, then there is only a single mode. For $\omega_p > \omega_T$, a new high-frequency mode appears whose phase and group velocities have opposite directions. However, such an anomalous dispersion persists as long as $\omega_p^2 < \omega_T^2 + \omega_L^2$ where ω_L is the frequency of the longitudinal magnons. The width of the gap in the region of anomalous dispersion exhibits a minimum equal to zero.
4. There is only one TE-type mode and, in contrast to the TM-type modes, the possible wave numbers of such a mode are bounded from above. The lower bound for the wave numbers and also the width and limits of the corresponding frequency range are very sensitive to the parameters of both the media in contact. When the c-axis is perpendicular to the interface, the TE mode has normal dispersion independent of the carrier density N. In the case when the c-axis is parallel to the interface, the dispersion of this mode is normal for $N < N_{cr}$ and anomalous for $N > N_{cr}$, where N_{cr} is half the critical value for the high-frequency TM mode.
5. The FTIR method for either Otto's or Kretschmann's geometry can be used to observe the properties of interface polaritons under study.
6. Under certain conditions, an electromagnetic wave incident obliquely on the two-layer structure consisting of an antiferromagnetic insulator and a semiconductor may be transmitted completely. This effect is caused by resonant excitation of the interface polaritons between the layers.

7. The zero-reflection transmission of the electromagnetic waves can take place for both s- and p-polarized waves and, in the first case, it is due to the presence of the optical magnons while in the second case it is due to the free carriers. In both cases, the effect can take place only in a well-defined region of carrier densities for fixed values of the wave frequency and angle of incidence depending on the orientation of the magnetic anisotropy axis $\boldsymbol{c}$, the ratio of layer thicknesses, permittivities, and other easily controlled parameters of the media. As a result, this effect could be used to obtain filters and polarizers with specified operating frequencies.

7.6 Influence of the Anisotropy of a Semiconductor Plasma on the Total Transmission Phenomenon

Let us now explore the influence of an anisotropy of the semiconductor parameters on the interface wave induced total transmission of infrared electromagnetic waves through the two-layered planar structure that consists of a uniaxial semiconductor (for example, Te or uniaxially deformed Si) and an antiferromagnet. We shall show that there is a considerable difference between the predictions of this model and those of the isotropic semiconductor model investigated in the previous sections. With the crystal axis in the xz-plane, making an angle φ with the normal to the interface of the sandwich structure, the semiconductor is described by the permittivity tensor given in Eqs. (5.1) and (5.1a). For the nonpolar crystal, the principal values of the tensor $\varepsilon_\perp(\omega)$ and $\varepsilon_{||}(\omega)$ are defined by Eqs. (1.30) and (1.31), where $\varepsilon_\perp^{\mathrm{L}}$ and $\varepsilon_{||}^{\mathrm{L}}$ are replaced by $\varepsilon_{\perp\infty}$ and $\varepsilon_{||\infty}$. The antiferromagnetic material is described by the permeability tensor (7.2) and the permittivity tensor defined by Eq. (7.6). The solution of the Maxwell equations for the waves with wave vector $\boldsymbol{k} = \left(k_{||}, 0, k_z\right)$ gives now two different waves for each medium. In the region occupied by the semiconductor, we obtain for the normal components of the wave vector

$$k_{\mathrm{oz}} = \mathrm{i}\sqrt{k_{||}^2 - \frac{\omega^2}{c^2}\varepsilon_\perp(\omega)} \tag{7.66}$$

and

$$k_{\mathrm{ez}} = k_{\mathrm{ez}}' + \mathrm{i}k_{\mathrm{ez}}'' = -k_{||}\frac{\varepsilon_{xz}}{\varepsilon_{zz}} + \mathrm{i}\sqrt{\frac{\varepsilon_\perp(\omega)\varepsilon_{||}(\omega)}{\varepsilon_{zz}^2}\left(k_{||}^2 - \frac{\omega^2}{c^2}\varepsilon_{zz}\right)} \tag{7.67}$$

which correspond to the TE- and TM-type waves, respectively. The corresponding expressions for the antiferromagnet are identical with those given by Eqs. (7.8) and (7.9).

Using the standard boundary conditions for the fields at the interface between two semi-infinite media, we obtain the dispersion relations for the coupled magnon–plasmon interface polaritons. For the TE-type waves it is practically identical with Eq. (7.14), where $\varepsilon(\omega)$ must be replaced by $\varepsilon_\perp(\omega)$ while for TM-type waves it is quite different from that given by Eq. (7.13) and has the form

$$\frac{c^2 k_{\|}^2}{\omega^2} = \frac{\varepsilon_{\|}\left[\varepsilon_\perp \varepsilon_{zz} - \mu_\perp(\omega)\varepsilon_{\|}(\omega)\varepsilon_\perp(\omega)\right]}{\varepsilon_{\|}\varepsilon_\perp - \varepsilon_{\|}(\omega)\varepsilon_\perp(\omega)} \tag{7.68}$$

This relation coincides with Eq. (7.13) only for the case of an isotropic semiconductor when one may substitute $\varepsilon_{zz} = \varepsilon_\perp(\omega) = \varepsilon_{\|}(\omega)$. The localization conditions for the waves described by Eq. (7.68) are

$$\varepsilon_\perp(\omega)\varepsilon_{\|}(\omega) < 0, \quad \varepsilon_{\|}\mu_\perp(\omega) < \frac{c^2 k_{\|}^2}{\omega^2} < \varepsilon_{zz} \tag{7.69}$$

The TE-type wave exists, as before, only in the frequency region $\omega_{\mathrm{T}} < \omega < \omega_{\mathrm{L}}$ while there are two existence regions for the TM-type waves:

$$\omega < \omega_{\mathrm{m}} \qquad \omega_\varphi < \omega < \omega_{\mathrm{M}} \tag{7.70}$$

where $\omega_{\mathrm{m}} = \min\{\omega_{0\perp}, \omega_{0\|}\}$, $\omega_{\mathrm{M}} = \max\{\omega_{0\perp}, \omega_{0\|}\}$ and

$$\omega_\varphi^2 = \frac{\varepsilon_{\perp\infty}\omega_{0\perp}^2 \sin^2\varphi + \varepsilon_{\|\infty}\omega_{0\|}^2 \cos^2\varphi}{\varepsilon_{\perp\infty}\sin^2\varphi + \varepsilon_{\|\infty}\cos^2\varphi} \tag{7.71}$$

In the following, we shall restrict ourselves to the case when the TM-type interface wave can be excited. Let us consider a sandwich structure made up of an antiferromagnetic layer of thickness d_2 and a uniaxial semiconductor layer of thickness d_1 A p-polarized plane electromagnetic wave is assumed to be incident upon this structure from the isotropic half-space with dielectric constant ε_{p}. Matching the solutions of the Maxwell equations at the boundary planes $z = d_1$, $z = 0$ and $z = -d_2$, after some simple algebra we obtain the coefficient of reflection in the form [76]

$$R_{\mathrm{p}} = \frac{G_1^2 + G_2^2}{G_3^2 + G_4^2} \tag{7.72}$$

where

$$G_1 = g_1(g_2^2 - g_3^2)\sinh\alpha \sinh\beta \tag{7.72a}$$

$$G_2 = g_2(g_1^2 + g_3^2)\cosh\alpha \sinh\beta + g_3(g_1^2 + g_2^2)\sinh\alpha \cosh\beta \tag{7.72b}$$

$$G_3 = g_1\left(g_2^2 + g_3^2\right)\sinh a \sinh\beta + 2g_1g_2g_3\cosh a\cosh\beta \tag{7.72c}$$

$$G_4 = g_2\left(g_1^2 - g_3^2\right)\cosh a\sinh\beta + g_3\left(g_1^2 - g_2^2\right)\sinh a\cosh\beta \tag{7.72d}$$

$$g_1 = \frac{\cos\theta}{\sqrt{\varepsilon_\mathrm{p}}} \qquad g_2 = \sqrt{\frac{\varepsilon_\mathrm{p}\sin^2\theta - \varepsilon_{zz}}{\varepsilon_\perp\left(\omega\varepsilon_{||}(\omega)\right)}} \qquad g_3 = \frac{\kappa}{\varepsilon_\perp} \tag{7.72e}$$

$$a = k_z'' d_1 \qquad \beta = \frac{\kappa\omega d_2}{c} \tag{7.72f}$$

Here θ is the angle of incidence and κ is given by Eq. (7.31).

Obviously, $R_\mathrm{p} = 0$ if the following conditions are fulfilled:

$$g_2 + g_3 = 0 \qquad a = \beta \tag{7.73}$$

But the first of them coincides exactly with the dispersion relation for the TM mode given by Eq. (7.68). Namely, the excitation of this wave provides the transmission of the incident wave's energy through the nontransparent two-layer structure. The second equation in Eq. (7.73) represents equality of effective thicknesses of the layers and provides energy balance: the intensity of the incident radiation is converted totally into that of the transmitted wave.

It is not difficult to see that the conditions (7.73) are satisfied simultaneously when the equality

$$-\frac{\varepsilon_\perp\varepsilon_{zz}}{\varepsilon_\perp(\omega)\varepsilon_{||}(\omega)} = \frac{d_1}{d_2} \tag{7.74}$$

holds. Using Eq. (7.74), one can obtain two different values for the frequencies of the total transmission $\Omega_i = \Omega_i(\varphi)$ and for the corresponding angles of incidence $\theta_i = \theta_i(\varphi)$, $i = 1, 2$:

$$\Omega_i^2 = \frac{2C}{B \pm \sqrt{B^2 - 4AC}} \tag{7.75}$$

$$\sin^2\theta_i = \frac{\varepsilon_{||}\varepsilon_{zz\infty}(d_1 + d_2)\left(\Omega_i^2 - \Omega_0^2\right)\left(\Omega_i^2 - \omega_\varphi^2\right)}{\varepsilon_\mathrm{p}\left(\Omega_i^2 - \omega_\mathrm{T}^2\right)\left[\Omega_i^2\left(d_1\varepsilon_{||} + d_2\varepsilon_{zz\infty}\right) - \omega_\varphi^2 d_2\varepsilon_{zz\infty}\right]} \tag{7.76}$$

where

$$A = \delta + \varepsilon_{zz\infty}, \quad B = \varepsilon_{zz\infty}\omega_\varphi^2 + \delta\left(\omega_{0||}^2 + \omega_{0\perp}^2\right), \quad C = \delta\omega_{0||}^2\omega_{0\perp}^2, \quad \delta = \frac{\varepsilon_{||\infty}\varepsilon_{\perp\infty}d_1}{\varepsilon_\perp d_2} \tag{7.77}$$

Here ω_φ is given by Eq. (7.71) and Ω_0 is the total transmission frequency for an s-polarized incident wave defined by Eq. (7.42).

Thus, a complete transmission of the incident radiation energy through the sandwich structure is due to the resonant excitation of the interface waves with frequencies Ω_i and tangential wave numbers

$$k_{\|}^{(i)} = \sqrt{\varepsilon_p}\frac{\Omega_i}{c}\sin\theta_i \quad i = 1, 2 \tag{7.78}$$

Unlike the isotropic semiconductor case, there are two different frequencies for tunneling of p-polarized waves, which depend on the orientation of the crystal axis with respect to the interface plane. If this axis is perpendicular ($\varphi = 0$) or parallel ($\varphi = \pi/2$) to the interface, there is only one such frequency given by

$$\Omega_\perp = \frac{\omega_{0\perp}}{\sqrt{1+\delta^{-1}\varepsilon_{\|\infty}}} \tag{7.79a}$$

for $\varphi = 0$, and

$$\Omega_\| = \frac{\omega_{0\|}}{\sqrt{1+\delta^{-1}\varepsilon_{\perp\infty}}} \tag{7.79b}$$

for $\varphi = \pi/2$. The doubling of the number of the complete transmission frequencies is the result of the semiconductor anisotropy and takes place only if $\varphi \neq 0, \pi/2$.

In Fig. 7.2, the frequencies Ω_1 and Ω_2 are shown schematically as functions of the angle φ. For $\omega_{0\|} > \omega_{0\perp}$ the frequency Ω_1 increases with increasing φ if $\frac{d_1}{d_2} > \gamma_1$ (see Fig. 7.2a) or $\gamma_1 > \frac{d_1}{d_2} > \gamma_2$ (Fig. 7.2b), where

$$\gamma_1 = \frac{\varepsilon_\perp m_\|}{\varepsilon_\perp m_\perp - \varepsilon_{\|\infty} m_\|}, \quad \gamma_2 = \frac{\varepsilon_{\perp\infty}\left(m_\| - m_\perp\right)}{\varepsilon_{\perp\infty} m_\perp - \varepsilon_{\|\infty} m_\|} \tag{7.80}$$

For $d_1/d_2 < \gamma_2$ (Fig. 7.2c), Ω_1 decreases with increasing φ. The upper frequency Ω_2 decreases in all these cases. For $\omega_{0\|} < \omega_{0\perp}$ the functions $\Omega_i(\varphi)$ are shown in Fig. 7.2d, e and f. In this case, Ω_2 increases with increasing φ while Ω_1 decreases if $d_1/d_2 > \gamma_1 m_\perp/m_\|$ (Fig. 7.2d) or $\gamma_1 m_\perp/m_\| > d_1/d_2 > \gamma_2$ (Fig. 7.2e), and increases if $d_1/d_2 < \gamma_2$ (Fig. 7.2f).

Hence, the values of Ω_i are very sensitive to the thickness ratio d_1/d_2 and the orientation of the crystal axis to the interface. This result is very important because it allows us to control the values of the total transmission frequencies by changing the ratio d_1/d_2 or the angle φ. On the other hand, it allows us to determine the angle φ by measuring the values of Ω_i.

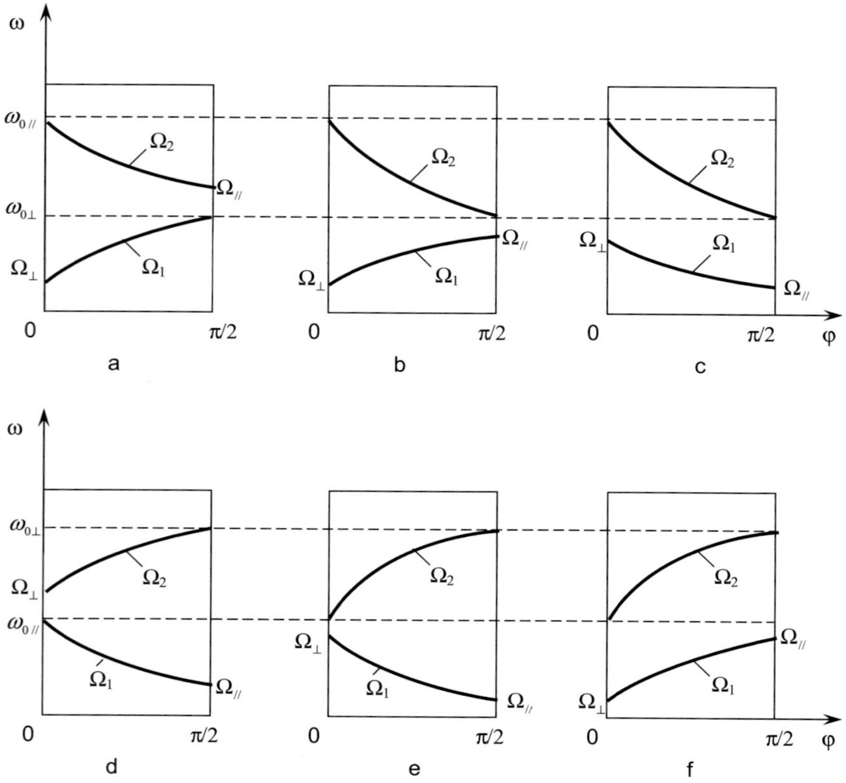

Fig. 7.2 Angle dependence of the total transmission frequencies Ω_1 and Ω_2.

8
Propagation of Electromagnetic Waves on a Lateral Surface of a Ferrite/Semiconductor Superlattice at Quantum Hall Effect Conditions

8.1
Model of Effective Permeability and Permittivity Tensors

In this chapter we continue our study of the coupled magnon–plasmon surface polaritons but, in contrast to the previous chapter, where the interface polaritons have been considered in a two-layered sandwich structure in the absence of an external magnetic field, here we will investigate the waves localized at the lateral surface of a superlattice which consists of alternating layers of semiconductor and ferromagnet, in the presence of a strong static magnetic field. Such an investigation requires a generalization of our results for the superlattices. The method of an effective homogeneous anisotropic medium will be used to find the permittivity and permeability tensors of the superlattice. This method can be used if the period of the superlattice is less than the wavelength of the polaritons [81]. The latter is also assumed to be long as compared with the cyclotron radius but less than the thickness of the superlattice plate. Each a medium in the presence of the magnetic field is described by a gyrotropic tensor. The effective anisotropic medium description allows us to express the permittivity and permeability tensors of the superlattice in terms of those for the individual layers.

Let us consider a superlattice as being infinite in the x- and z-axis directions and filling the space region $y<0$ (Fig. 8.1). Medium 1 is a semiconductor layer of thickness d_1 and medium 2 is a ferrite layer of thickness d_2. The layers are parallel to the (x, y)-plane and are perpendicular to the static magnetic field: $\boldsymbol{B}_0 || oz$. An individual ferrite layer is described by the scalar static dielectric constant ε_F and the magnetic permeability tensor

$$\boldsymbol{\mu}(\omega) = \begin{pmatrix} \mu_1 & i\mu_2 & 0 \\ -i\mu_2 & \mu_1 & 0 \\ 0 & 0 & 1 \end{pmatrix} \tag{8.1}$$

where

$$\mu_1 = \frac{\omega^2 - \omega_H(\omega_H + \omega_M)}{\omega^2 - \omega_H^2} \tag{8.1a}$$

Radiowaves and Polaritons in Anisotropic Media. Roland H. Tarkhanyan and Nikolaos K. Uzunoglu

ISBN: 3-527-40615-8

$$\mu_2 = \frac{\omega\omega_M}{\omega^2 - \omega_H^2} \tag{8.1b}$$

Here $\omega_H = B_0\gamma, \omega_M = M_s\gamma$; γ and M_s denote the gyromagnetic ratio and the saturation magnetization, respectively. It is evident that Eqs. (8.1a) and (8.1b) are valid only if the dissipation of the waves can be neglected.

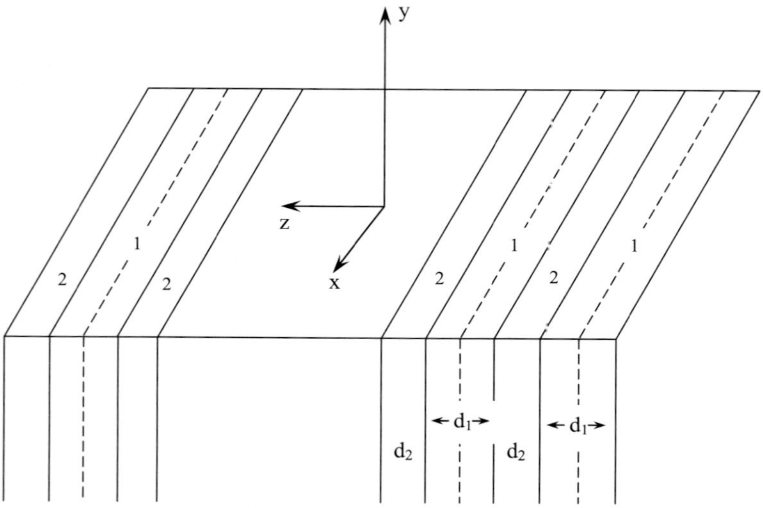

Fig. 8.1 Geometry of the ferrite/semiconductor superlattice in the external static magnetic field. Dashed lines denote 2D electron gas planes.

An individual semiconductor layer is described by the magnetic permeability $\mu = 1$ and by the dielectric permittivity tensor

$$\varepsilon(\omega) = \begin{pmatrix} \varepsilon_1 & i\varepsilon_2 & 0 \\ -i\varepsilon_2 & \varepsilon_1 & 0 \\ 0 & 0 & \varepsilon_3 \end{pmatrix} \tag{8.2}$$

The expressions for ε_1, ε_2 and ε_3 have quite different forms at the quantum Hall effect conditions and in the case of a nonquantizing magnetic field. In the latter case they are given by Eqs. (6.43) and (6.44). In the first case we assume that the semiconductor layers are a GaAs–AlGaAs-type quantum well system and a two-dimensional (2D) electron gas plane is described by the local conductivity tensor [82]

$$\boldsymbol{\sigma} = \begin{pmatrix} \sigma_{xx} & \sigma_{xy} \\ -\sigma_{xy} & \sigma_{xx} \end{pmatrix} \tag{8.3}$$

In the quantum Hall effect conditions [83, 84]

$$\sigma_{xx} = 0 \qquad \sigma_{xy} = -\frac{e^2}{h}s \tag{8.4}$$

where the integer s is the number of the filling Landau levels and h is the Planck constant. Then, an individual semiconductor layer of thickness d_1 can be described by the complex permittivity tensor (8.2), where $\varepsilon_1 = \varepsilon_3 \equiv \varepsilon_L$ is the average static dielectric constant and

$$\varepsilon_2 = \frac{\sigma_{xy}}{\varepsilon_0 \omega d_1} \tag{8.5}$$

In the case when the period of the superlattice $d = d_1 + d_2$ is less than the wavelength $\lambda_z = 2\pi/k_z$ one may average Eqs. (8.1) and (8.2) and obtain, with the accuracy of order $d/\lambda_z << 1$, the effective permeability and permittivity tensors in the form

$$\boldsymbol{\mu}_{eff} = \bar{\mu}\begin{pmatrix} 1 & i\varsigma & 0 \\ -i\varsigma & 1 & 0 \\ 0 & 0 & \bar{\mu}^{-1} \end{pmatrix} \tag{8.6}$$

$$\boldsymbol{\varepsilon}_{eff} = \bar{\varepsilon}\begin{pmatrix} 1 & i\nu & 0 \\ -i\nu & 1 & 0 \\ 0 & 0 & \xi \end{pmatrix} \tag{8.7}$$

where

$$\bar{\mu} = \frac{1}{d}(d_1 + \mu_1 d_2) = 1 - \frac{\omega_H \omega_1}{\omega^2 - \omega_H^2} \qquad \omega_1 \equiv \omega_M \frac{d_2}{d} \tag{8.8}$$

$$\varsigma = \frac{\mu_2 d_2}{\bar{\mu} d} = \frac{\omega_1}{\omega^2 - \omega_H(\omega_H + \omega_1)} \tag{8.9}$$

$$\bar{\varepsilon} = \frac{1}{d}(\varepsilon_L d_1 + \varepsilon_F d_2) \tag{8.10}$$

$$\nu = -\frac{\varepsilon_2 d_1}{\bar{\varepsilon} d} \tag{8.11}$$

$$\xi = \frac{1}{\bar{\varepsilon} d}(\varepsilon_3 d_1 + \varepsilon_F d_2) \tag{8.12}$$

At quantum Hall effect conditions, which are the main subject of sections to come, Eq. (8.12) gives $\xi = 1$. Also, using Eqs. (8.4), (8.5) and (8.11), we may write ν as

$$\nu = s\frac{\nu_0}{\omega} \qquad \nu_0 = \frac{e^2}{h\varepsilon_0\bar{\varepsilon}d} \qquad s = 1,\ 2, \ldots \tag{8.13}$$

Thus, the ferrite/semiconductor superlattice in the effective homogeneous anisotropic medium method is really a bigyrotropic medium, which is described by the magnetic permeability and dielectric permittivity tensors given by Eqs. (8.6) and (8.7).

8.2
Partial Waves and Electromagnetic Field Structure

Before obtaining the dispersion relations for the nonradiative surface waves propagating along the lateral surface of the superlattice $y = 0$, let us consider the collective modes which arise from the coupling between the magnons and 2D magnetoplasmons in the presence of fluctuating electromagnetic fields at quantum Hall effect conditions.

First of all, let us introduce some new notation:

$$\varepsilon_\nu = \bar{\varepsilon}(1-\nu^2) \qquad \mu_\nu = \bar{\mu}(1-\varsigma^2) \qquad \delta = \varsigma - \nu \tag{8.14}$$

$$n_x = \frac{ck_x}{\omega} \qquad n_z = \frac{ck_z}{\omega} \qquad \chi = \mathrm{i}\frac{ck_y}{\omega} \tag{8.15}$$

$$a = n_z^2 - \bar{\varepsilon}\mu_\nu \qquad \beta = \frac{n_z^2}{\bar{\mu}} - \varepsilon_\nu \tag{8.16}$$

Then the Maxwell equations

$$\mathrm{rot}\boldsymbol{H} = -\mathrm{i}\omega\varepsilon_0\boldsymbol{\varepsilon}_{\mathit{eff}}\boldsymbol{E} \quad \mathrm{rot}\boldsymbol{E} = \mathrm{i}\omega\mu_0\boldsymbol{\mu}_{\mathit{eff}}\boldsymbol{H} \tag{8.17}$$

for the plane waves $\boldsymbol{E}, \boldsymbol{H} \sim \exp[\mathrm{i}(\boldsymbol{kr} - \omega t)]$ give

$$a_{ij}E_j = 0 \tag{8.18}$$

where the tensor $\boldsymbol{a}$ is given by

$$a = \begin{pmatrix} a - \mu_\nu\chi^2 & -\mathrm{i}(a\nu + \delta n_z^2 - \chi\mu_\nu n_x) & n_z(n_x - \chi\varsigma) \\ \mathrm{i}(a\nu + \delta n_z^2 + \chi\mu_\nu n_x) & a + \mu_\nu\chi^2 & \mathrm{i}n_z(\chi - \varsigma n_x) \\ n_z(n_x + \chi\varsigma) & -\mathrm{i}n_z(\chi + \varsigma n_x) & \chi^2 - n_x^2 + \bar{\varepsilon}\mu_\nu \end{pmatrix} \tag{8.19}$$

or

$$b_{ij}H_j = 0 \tag{8.20}$$

where

$$b = \begin{pmatrix} -\beta\bar{\mu} + (1-\nu^2)\chi^2 & -\mathrm{i}[(1-\nu^2)\chi n_x + \bar{\mu}(\beta\nu - \delta\varepsilon_\nu)] & n_z(n_x + \chi\nu) \\ \mathrm{i}[(1-\nu^2)\chi n_x - \bar{\mu}(\beta\nu - \delta\varepsilon_\nu)] & \beta\bar{\mu} + (1-\nu^2)n_x^2 & \mathrm{i}n_z(\chi + \nu n_x) \\ n_z(n_x - \chi\nu) & -\mathrm{i}n_z(\chi - \nu n_x) & \chi^2 - n_x^2 + \varepsilon_\nu \end{pmatrix} \tag{8.21}$$

Setting det $\boldsymbol{a} = 0$ or det $\boldsymbol{b} = 0$, we obtain two partial waves with different values for χ at any given values for the tangential components k_x and k_z of the wave vector:

$$\chi_{1,2}^2 = n_x^2 + \frac{1}{2}\left\{\alpha + \beta \mp \left[(-\beta)^2 + 4\bar{\varepsilon}\delta^2 n_z^2\right]^{1/2}\right\} \tag{8.23}$$

Thus, the electromagnetic field in the space region $y < 0$ can be written as a superposition of the partial waves with different polarizations:

$$\boldsymbol{E}(\boldsymbol{r}, t) = \boldsymbol{E}(y)\exp[i(k_x x + k_z z - \omega t)]$$

where

$$\boldsymbol{E}(y) = \boldsymbol{E}_1\exp(\omega\chi_1 y/c) + \boldsymbol{E}_2\exp(\omega\chi_2 y/c) \tag{8.24}$$

and an analogous expression for the magnetic field vector $\boldsymbol{H}(\boldsymbol{r}, t)$. To find the polarization of the partial waves, let us use Eqs. (8.20) and (8.21). The field structure of the wave with χ_1 can be described as

$$\boldsymbol{H}_1 = \left\{h_{1x}, \mathrm{i}h_{1y}, 1\right\}H_{1z} \tag{8.25}$$

$$\boldsymbol{E}_1 = \left\{\mathrm{i}e_{1x}, e_{1y}, -\mathrm{i}e_{1z}\right\}\varepsilon_\nu^{-1}H_{1z} \tag{8.26}$$

where

$$\boldsymbol{e}_1 = \begin{pmatrix} e_{1x} \\ e_{1y} \\ e_{1z} \end{pmatrix} = \begin{pmatrix} -\nu n_z & n_z & \chi_1 + \nu n_x \\ -n_z & \nu n_z & n_x + \nu\chi_1 \\ \varepsilon_\nu\bar{\varepsilon}^{-1}\chi_1 & \varepsilon_\nu\bar{\varepsilon}^{-1}n_x & 0 \end{pmatrix}\begin{pmatrix} h_{1x} \\ h_{1y} \\ 1 \end{pmatrix} \tag{8.27}$$

$$h_{1x} = \frac{B_{31}}{B_{33}} = \frac{B_{11}}{B_{13}} \qquad h_{1y} = -\mathrm{i}\frac{B_{32}}{B_{33}} = -\mathrm{i}\frac{B_{12}}{B_{13}} \tag{8.28}$$

Here $B_{ij} = (-1)^{i+j} x$ minor of the tensor $\boldsymbol{b}$ corresponding to the element b_{ij}, that is,

$$B_{11} = b_{22}b_{33} - b_{23}b_{32} = \varepsilon_\nu\left[n_x^2 n_z^2(1 - \bar{\mu}^{-1}) + (n_x^2 - \bar{\varepsilon}\bar{\mu})(\beta + n_x^2 - \chi_1^2)\right] \tag{8.29}$$

$$B_{12} = b_{23}b_{31} - b_{21}b_{33} = i\varepsilon_v\left[(\beta + n_x^2 - \chi_1^2)(\chi_1 n_x + \varsigma\varepsilon\mu) - \delta n_z^2\varepsilon + n_x n_z^2\chi_1(1-\mu^{-1})\right] \tag{8.30}$$

$$B_{13} = b_{21}b_{32} - b_{22}b_{31} = \varepsilon_v n_z\left[n_x(a + n_x^2 - \chi_1^2) - \delta\bar{\varepsilon}\bar{\mu}(\chi_1 + \varsigma n_x)\right] \tag{8.31}$$

Analogously, using Eqs. (8.18) and (8.19), for the fields of the partial waves described by χ_2 we obtain

$$\boldsymbol{E}_2 = \left\{e_{2x}, ie_{2y}, 1\right\}E_{2z} \tag{8.32}$$

$$\boldsymbol{H}_2 = \left\{-ih_{2x}, h_{2y}ih_{2z}\right\}\mu_v^{-1}E_{2z} \tag{8.33}$$

where

$$\boldsymbol{h}_2 = \begin{pmatrix} h_{2x} \\ h_{2y} \\ h_{2z} \end{pmatrix} = \begin{pmatrix} \varsigma n_z & n_z & \chi_2 - \varsigma n_x \\ n_z & \varsigma n_z & \varsigma\chi_2 - n_x \\ \mu_v\chi_2 & \mu_v n_x & 0 \end{pmatrix} \begin{pmatrix} e_{2x} \\ e_{2y} \\ 1 \end{pmatrix} \tag{8.34}$$

$$e_{2x} = \frac{A_{31}}{A_{33}} = \frac{A_{11}}{A_{13}} \qquad e_{2y} = -i\frac{A_{32}}{A_{33}} = -i\frac{A_{12}}{A_{13}} \tag{8.35}$$

and $A_{ij} = (-1)^{i+j}x$ minor of the tensor $\boldsymbol{a}$ for the element a_{ij}. Note that the polarization parameters $e_{1i}, h_{2i}, i = x, y, z$ and $h_{1j}, e_{2j}, j = x. y$, are real quantities.

To simplify all the above expressions, let us consider two special cases for which the wave given by Eq. (8.24) splits in two.

A. $k_z = 0$ (the Voigt configuration).

In this case there are two independent waves propagating in the direction perpendicular to the static magnetic field: a TM-polarized wave described by the relations

$$\chi_1^2 = n_x^2 - \varepsilon_v \tag{8.36}$$

$$\boldsymbol{H}_1 = \{0, 0, H_{1z}\} \quad \boldsymbol{E}_1 = \{i(\chi_1 + \nu n_x), n_x + \chi_1\nu, 0\}\varepsilon_v^{-1}H_{1z} \tag{8.37}$$

and a TE-polarized wave, for which

$$\chi_2^2 = n_x^2 - \bar{\varepsilon}\mu_v \tag{8.38}$$

$$\boldsymbol{E}_2 = \{0, 0, E_{2z}\} \qquad \boldsymbol{H}_2 = \{i(\varsigma n_x - \chi_2), \varsigma\chi_2 - n_x, 0\}\mu_v^{-1}E_{2z} \tag{8.39}$$

B. $\delta = 0$

In this special case $a = \beta\bar{\mu}$, $\chi_2^2 - \chi_1^2 = a - \beta$, and the frequency of the waves is given by

$$\omega^2 = \frac{s\nu_0\omega_H(\omega_H + \omega_1)}{s\nu_0 - \omega_1} \tag{8.40}$$

There are two independent waves with discrete frequencies: an extraordinary wave for which

$$\chi_1^2 = n_x^2 - \varepsilon_\nu + n_z^2\bar{\mu}^{-1} \tag{8.41}$$

$$\boldsymbol{H}_1 = \{n_x n_z a^{-1}, -\mathrm{i}\chi_1 n_z a^{-1}, 1\} H_{1z} \tag{8.42}$$

$$\boldsymbol{E}_1 = -\{\mathrm{i}(\nu n_x + \chi_1),\ n_x + \nu\chi_1, 0\}\beta^{-1} H_{1z} \tag{8.43}$$

and an ordinary wave described by the relations

$$\chi_2^2 = n_x^2 + n_z^2 - \bar{\varepsilon}\mu_\nu \tag{8.44}$$

$$\boldsymbol{E}_2 = \{n_x n_z a^{-1}, -\mathrm{i}\chi_2 n_z a^{-1}, 1\} E_{2z} \tag{8.45}$$

$$\boldsymbol{H}_2 = \{\mathrm{i}(\chi_2 - \nu n_x),\ n_x - \nu\chi_2, 0\}\bar{\varepsilon} a^{-1} E_{2z} \tag{8.46}$$

Note that the parameters μ_ν, $\bar{\mu}$ and $\bar{\varepsilon}_\nu$ in Eqs. (8.41)–(8.46) have discrete values:

$$\bar{\mu} = \frac{\omega_H + \omega_1}{s\nu_0 + \omega_H} \tag{8.47}$$

$$\mu_\nu = 1 + \frac{\omega_1 - s\nu_0}{\omega_H} \tag{8.48}$$

$$\varepsilon_\nu = \bar{\varepsilon}\left[1 + \frac{s\nu_0(\omega_1 - s\nu_0)}{\omega_H(\omega_H + \omega_1)}\right] \tag{8.49}$$

Both the ordinary and extraordinary waves are elliptically polarized. The waves can be described by the refractive indices

$$n_o^2 = \bar{\varepsilon}\mu_\nu \tag{8.50}$$

and

$$n_e^2 = \frac{\bar{\varepsilon}\mu_\nu}{1 + (\bar{\mu} - 1)\sin^2\theta} \tag{8.51}$$

respectively, where θ is the angle between the wave vector and the axis of the superlattice. The existence of the bulk waves with discrete frequencies and discrete refractive indices has been first mentioned in Ref. [85]. The waves exist only if the condition

$$\omega_1 < sv_0 < \omega_1 + \omega_H \tag{8.52}$$

is fulfilled. For the existence of the extraordinary wave the additional condition

$$\sin\theta < \sqrt{\frac{sv_0 + \omega_H}{sv_0 - \omega_1}} \tag{8.53}$$

must be fulfilled, too.

8.3
Interface Waves Propagating Along the Lateral Surface

Let us assume that the space region $y > 0$ is filled by an isotropic insulator of dielectric constant ε_d (for the vacuum $\varepsilon_d = 1$). A plane electromagnetic wave in this region can be written as

$$\boldsymbol{E}_d, \boldsymbol{H}_d \sim \exp[\mathrm{i}(k_x x + k_z z - \omega t) - \omega \chi_d y/c] \tag{8.54}$$

where

$$\chi_d^2 = n_x^2 + n_z^2 - \varepsilon_d \tag{8.55}$$

Excluding normal components of the field vectors, for the tangential components from the Maxwell equations we obtain

$$\begin{pmatrix} E_{dx} \\ E_{dz} \end{pmatrix} = \frac{\mathrm{i}}{\varepsilon_d \chi_d} \begin{pmatrix} n_x n_z & \varepsilon_d - n_x^2 \\ n_z^2 - \varepsilon_d & -n_x n_z \end{pmatrix} \begin{pmatrix} H_{dx} \\ H_{dz} \end{pmatrix} \tag{8.56}$$

Using the standard boundary conditions of continuity for the tangential components of the fields, and Eqs. (8.24), (8.26), (8.33), (8.54) and (8.56), we obtain a set of four linear homogeneous algebraic equations for the four unknowns H_{1z}, E_{2z}, H_{dx} and H_{dz}. The condition of existence of nontrivial solutions for the system of equations gives the dispersion relation for the interface waves, in the form

$$F_1 G_2 - n_z^2 F_2 G_1 = 0 \tag{8.57}$$

where

$$F_1 = (a + \gamma_1)f_1 + n_x n_z^2 \delta g_1 \qquad F_2 = f_2\delta + n_x(\beta + \gamma_2)g_2 \tag{8.57a}$$

$$G_1 = \delta\bar{\varepsilon}p_1 - n_x q_1(a + \gamma_1) \qquad G_2 = p_2(\beta + \gamma_2) - n_x n_z^2 q_2 \delta \tag{8.57b}$$

$$f_j = \bar{\mu}^{-1} n_x^2 n_z^2 + \gamma_j(n_x^2 - \varepsilon_\mathrm{d}) + \varepsilon_\mathrm{d}\chi_\mathrm{d}(\nu n_x + \chi_j) \quad j = 1, 2. \tag{8.57c}$$

$$g_j = \rho\chi_\mathrm{d} + \chi_j - \varsigma n_x \quad \rho = \varepsilon_\mathrm{d}\bar{\varepsilon}^{-1} < 1 \quad \gamma_j = n_x^2 - \chi_j^2 \tag{8.57d}$$

$$p_j = \rho\chi_\mathrm{d}\gamma_j + \left(\chi_j - \varsigma n_x\right)\left(n_x^2 - \chi_\mathrm{d}^2\right) \qquad q_j = \gamma_j - \bar{\mu}^{-1}\left(n_x^2 - \chi_\mathrm{d}^2\right) \tag{8.57e}$$

Using Eq. (8.23), after some simple algebra Eq. (8.57) can be written as

$$A_1 + n_x(A_2 + A_3 + A_4) + \chi_\mathrm{d}A_5 + n_x^2(\chi_1 + \chi_2)A_6 = 0 \tag{8.58}$$

where

$$A_1 \equiv \left(a\beta - \bar{\varepsilon}\delta^2 n_z^2\right)\left[\varepsilon_\mathrm{d}(\chi_1 + \chi_2 + \chi_\mathrm{d}) + \bar{\varepsilon}\chi_\mathrm{d}\right] \tag{8.58a}$$

$$A_2 \equiv \delta\bar{\varepsilon}\left(n_z^2 + \varepsilon_\mathrm{d}\right)\left(n_x^2 + \chi_1\chi_2\right) \tag{8.58b}$$

$$A_3 \equiv \bar{\varepsilon}\left(n_z^2 - \varepsilon_\mathrm{d}\right)\left[a\varsigma - \beta\nu + \delta(\varepsilon + \bar{\varepsilon}\mu_\nu)\right] \tag{8.58c}$$

$$A_4 \equiv \chi_\mathrm{d}(\chi_1 + \chi_2)\left[\varepsilon_\mathrm{d}\left(a\nu + \delta n_z^2\right) + \bar{\varepsilon}(\delta\varepsilon_\nu - \beta\nu)\right] \tag{8.58d}$$

$$A_5 \equiv \left(n_x^2 + \chi_1\chi_2\right)\left(\bar{\varepsilon}\beta + \varepsilon_\mathrm{d}a\right) \tag{8.58e}$$

$$A_6 \equiv \left(\varsigma\nu n_x^2 - \chi_1\chi_2\right)\left(\varepsilon_\mathrm{d}n_z^2\bar{\mu}^{-1} - \bar{\varepsilon}\varepsilon_\nu\mu_\nu\right) \tag{8.58f}$$

As we know, the nonradiative interface waves exist only if χ_1, χ_2 and χ_d are real and positive. It is possible if the following conditions are fulfilled, simultaneously:

$$n_x^2 + n_z^2 > \varepsilon_\mathrm{d} \tag{8.59}$$

$$n_x^2 + a > 0 \quad n_x^2 + \beta > 0 \tag{8.60}$$

$$\left(n_x^2 + a\right)\left(n_x^2 + \beta\right) > \bar{\varepsilon}\delta^2 n_z^2 \tag{8.61}$$

Solutions of Eq. (8.58) give the frequency of the interface waves as a nonexplicit function of the tangential wave numbers k_x and k_z at a given value of the quantizing magnetic field. To find an analytical expression for the function $\omega(k_x, k_z)$ one must proceed by an approximation. But, above all, let us make some conclusions on the basis of the general dispersion relation (8.58) and the localization conditions (8.59)–(8.61).

It is evident that solutions of Eq. (8.58) correspond to mixed spin–electromagnetic interface waves with quantizing frequencies due to the terms containing $\nu \sim$

s [see Eq. (8.13)]. Besides, we conclude that there is a nonreciprocity between the left ($k_x < 0$) and right ($k_x > 0$) directions of propagation due to the terms $\sim n_x$. The reciprocity can be realized only if $k_x = 0$ i.e. only for the Faraday configuration. In this case, Eqs. (8.58)–(8.61) and (8.23) give

$$(\chi_1 + \chi_2)(\chi_\mathrm{d}^2 + \bar{\varepsilon}^{-1}\varepsilon_\mathrm{d}\chi_1\chi_2) + \chi_\mathrm{d}\left[\beta + \chi_1\chi_2 + \bar{\varepsilon}^{-1}\varepsilon_\mathrm{d}(\alpha + \chi_1\chi_2)\right] = 0 \tag{8.62}$$

and

$$\alpha < 0, \quad \beta > 0 \tag{8.63}$$

It is not difficult to see that the left-hand side of Eq. (8.62) is a real and positive quantity, i.e. Eq. (8.62) has no solution. This means that there is no nonradiative interface wave in the case of the Faraday configuration.

Hence, the interface waves exist only if $n_x \neq 0$. Also, we conclude that the nonreciprocity is one of the most characteristic properties of the surface electromagnetic waves in the presence of a static magnetic field.

8.4 Spectrum of Interface Modes for the Voigt Configuration

In this particular case, $k_z = 0$ and Eq. (8.58) splits in two:

$$\varepsilon_\mathrm{d}(\chi_1 + \nu n_x) + \varepsilon_\nu\chi_\mathrm{d} = 0 \tag{8.64}$$

$$\chi_2 - n_x\varsigma + \mu_\nu\chi_\mathrm{d} = 0 \tag{8.65}$$

where χ_1 and χ_2 are given by Eqs. (8.36) and (8.38), and $\chi_\mathrm{d}^2 = n_x^2 - \varepsilon_\mathrm{d}$. Equation (8.64) corresponds to TM-type waves and Eq. (8.65) to TE-type waves. It is evident that TM modes are only sensitive to the parameters of the semiconductor layers, and correspond to the interface plasmon–polaritons with a discrete spectrum, while TE modes are only sensitive to the parameters of the ferrite layers, and correspond to the interface magnon–polaritons with a classical spectrum. We have to note that the mixed interface waves arising due to the coupling of magnetoplasmons, magnons and pure electromagnetic waves can only exist if both k_x and k_z are different from zero. Also, we note that the refractive indices for TM waves are quantized:

$$n_\pm^2 = \frac{\varepsilon_\mathrm{d}}{\omega^2 - \Omega_+^2}\left[\frac{\bar{\varepsilon}\omega^2}{\bar{\varepsilon} + \varepsilon_\mathrm{d}} - \Omega_+^2 \pm \frac{2\varepsilon_\mathrm{d}\Omega_-\omega^2}{(\bar{\varepsilon} + \varepsilon_\mathrm{d})^2}\sqrt{\frac{\bar{\varepsilon}\varepsilon_\mathrm{d}}{\Omega_-^2 - \omega^2}}\right] \tag{8.66}$$

where

$$\Omega_{\pm} = \frac{sv_0\bar{\varepsilon}}{\bar{\varepsilon} \pm \varepsilon_{\rm d}} \tag{8.67}$$

Thus, there are two TM-type waves, one of which has a resonance ($n_{-} \to \infty$) at $\omega = \Omega_{+}$ while the second is at $\omega = \Omega_{-}$. The existence region of these waves for a given value of the integer "s" is shown in Fig. 8.2, in the $\{\omega^2, c^2k_x^2/\varepsilon_{\rm d}\}$-plane. On the left, the existence region is bounded by the light line (1) described by

$$\omega^2 = \frac{c^2k_x^2}{\varepsilon_{\rm d}} \tag{8.68}$$

and by the straight line (2) for which

$$\omega^2 = s^2v_0^2 + \frac{c^2k_x^2}{\bar{\varepsilon}} \tag{8.69}$$

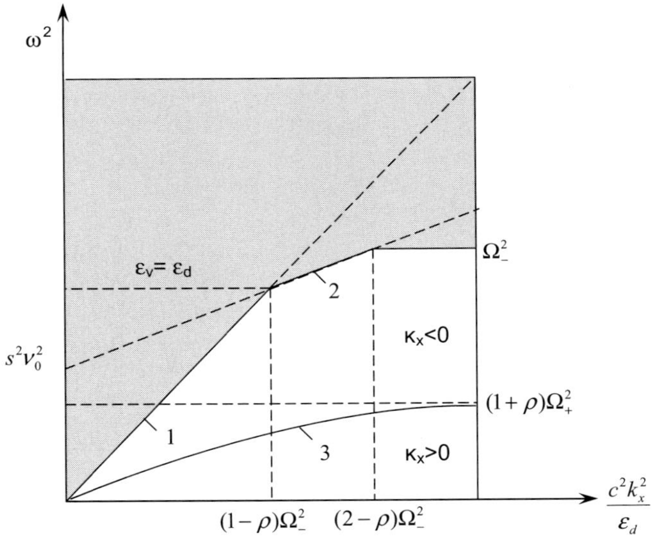

Fig. 8.2 The existence region for the interface TM waves. The shaded region is forbidden for the propagation.

The value of the frequency

$$\omega = sv_0\sqrt{\frac{\bar{\varepsilon}}{\bar{\varepsilon} - \varepsilon_{\rm d}}} \tag{8.70}$$

at which the lines (1) and (2) intersect corresponds to the solution of the equation

$$\varepsilon_v(\omega) = \varepsilon_{\rm d} \tag{8.71}$$

where ε_v is given by Eq. (8.14). This value of ω as well as the upper boundary of the transparency region $\omega = \Omega_-$ jumps up on increasing the integer s. Curve (3) in Fig. 8.2 is described by

$$\frac{c^2 k_x^2}{\varepsilon_d} = \frac{\omega^2 (s^2 v_0^2 - \omega^2)}{s^2 v_0^2 - \omega^2 (1 + \bar{\varepsilon}^{-1} \varepsilon_d)} \tag{8.72}$$

It separates the regions of propagation in the opposite directions perpendicular to the magnetic field: $k_x < 0$ above and $k_x > 0$ under the curve (3).

The spectrum of TM modes is given schematically in Fig. 8.3, for the first three values of the integer s. The low-frequency modes described by the refractive index n_- can only be propagated to the right ($k_x > 0$) while the high-frequency modes, for which $n = n_+$, propagate only to the left. The frequency increases with increasing $|k_x|$ for both TM modes. At a given value of s, the frequency of the lower branches rises from zero at $k_x = 0$ to the asymptotic value $\Omega_+(s)$ in the nonretardation limit $ck_x \to \infty$. Each of the high-frequency branches exists in the interval

$$\Omega_+(s) < \omega < \Omega_-(s) \tag{8.73}$$

and in the wave-number region for which

$$|k_x| > \sqrt{\varepsilon_d} \frac{\Omega_+(s)}{c} \tag{8.74}$$

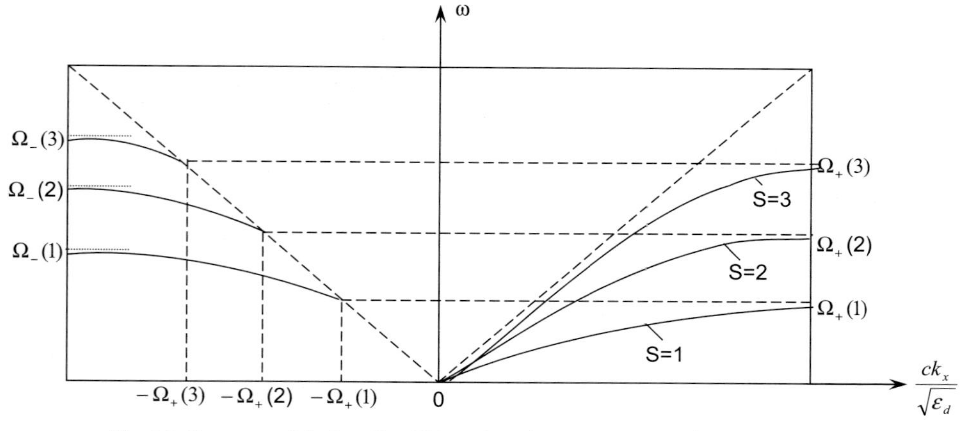

Fig. 8.3 Spectrum of the interface TM modes (dispersion curves) for $s = 1, 2, 3$ at $\rho < 0.2$.

8.5
Interface Magnon–Plasmon Polaritons in Some Particular Cases

As we have seen in previous sections, the coupled interface magnon–plasmon polaritons do not exist in both the Faraday and the Voigt configurations, that is they propagate only if $k_x \neq 0$ and $k_z \neq 0$. In this case, the mixed waves can be considered analytically only when the simplifying condition $\delta = 0$ or $\varsigma = \nu$ is fulfilled. Then, from the general dispersion relation (8.57), we obtain

$$f_1 p_2 + n_x^2 n_z^2 q_1 g_2 = 0 \tag{8.75}$$

which corresponds to the interface waves with discrete frequencies defined in Eq. (8.40). In contrast to the independent partial waves described by Eqs. (8.40)–(8.52), the wave (8.75) really is a superposition of the ordinary and extraordinary waves in the space region $y < 0$. There is some restriction for the angle θ between the static magnetic field and the tangential wave vector $\boldsymbol{k}_{\|}$. To see that, let us consider the particular case when the additional condition

$$q_1 = \frac{\bar{\varepsilon}}{\bar{\mu}}(\mu_\nu - \rho) = 0 \tag{8.76}$$

is fulfilled. This is possible only if

$$s\nu_0 = \omega_1 + (1 - \rho)\omega_{\mathrm{H}} \tag{8.77a}$$

or

$$\omega^2 = (\omega_1 + \omega_{\mathrm{H}})\left(\frac{\omega_1}{1 - \rho} + \omega_{\mathrm{H}}\right) \tag{8.77b}$$

Then $\chi_2 = \chi_{\mathrm{d}}$ and Eq. (8.75) splits in two:

$$(1 + \rho)\chi_{\mathrm{d}} = \nu n_x \tag{8.78}$$

and

$$\chi_1 + \frac{\chi_{\mathrm{d}}}{\bar{\mu}} = -\nu n_x \tag{8.79}$$

It means that there are two independent surface waves with the same frequency (8.77b) but of different wave numbers. The wave described by Eq. (8.78) exists only if $k_x > 0$ while the wave (8.79) propagates only to the left ($k_x < 0$). Both equations (8.78) and (8.79) can be written as

$$\frac{c^2 k_{\parallel}^2}{\omega^2} = \frac{\varepsilon_{\mathrm{d}}}{1 - a_i \sin^2\theta} \tag{8.80}$$

where

$$a_1 = \frac{s v_0 (1-\rho)}{(1+\rho)^2 (\omega_1 + \omega_{\mathrm{H}})} \tag{8.81}$$

for the wave (8.78) and

$$a_2 = \frac{(\omega_1 + \omega_{\mathrm{H}})\left(s v_0 + \omega_1 + 2\sqrt{s v_0 \omega_1}\right)}{\omega_{\mathrm{H}}^2 (1-\rho)} \tag{8.82}$$

for the wave (8.79). Using Eq. (8.77a), one can see that $a_1 < 1$ and there is no restriction for the angle θ, while $a_2 > 1$ and therefore the wave (8.79) can be propagated only in the directions for which

$$\sin\theta < \frac{1}{\sqrt{a_2}}. \tag{8.83}$$

Part 3
Electromagnetic Instabilities in Uniaxial Semiconductors with Hot Carriers

Roland H. Tarkhanyan

Radiowaves and Polaritons in Anisotropic Media. Roland H. Tarkhanyan and Nikolaos K. Uzunoglu

ISBN: 3-527-40615-8

Introduction

Since the invention of the tunnel diode and the discovery of current microwave oscillations in n-GaAs [86], the electrical instabilities in p–n junctions and bulk semiconductors with a negative differential resistance (NDR) have attracted much interest (see, for example, Refs. [87–90]). It is well known that NDR devices are used as microwave generators to convert dc power into ac power [91, 92]. In contrast to the Esaki tunnel diode, which is used for low-power applications since the voltage range in which NDR occurs is small, the Gunn diode can be used when higher power output is required.

Gallium arsenide exhibits NDR because of its band-structure properties due to the transfer of electrons in reciprocal lattice vector space. The Gunn effect was theoretically predicted by Ridley and Watkins [93] and Hilsum [94]. It is connected with the formation of a high potential electric field domain at the negative contact. When the domain moves through the semiconductor with the high field drift velocity and vanishes at the positive contact, the process starts again and leads to the current microwave oscillations.

Some time afterwards, it was also shown that the real space transfer effect can give rise to NDR in layered semiconductor heterostructures [95]. The emission of electromagnetic waves at microwave frequencies from Ge and GaAs crystals with NDR in the absence of high-field domains has been reported by Baynham [96, 97]. Excitation and amplification of the waves in isotropic (cubic) materials with NDR was studied theoretically in Refs. [98, 99]. In contrast to longitudinal oscillations of the space charge in the Gunn effect, the transverse electromagnetic waves travel in the direction perpendicular to the static electric field $\boldsymbol{E}_0$ and are polarized so that their magnetic field is perpendicular to $\boldsymbol{E}_0$.

This part of the present book is concerned with the electromagnetic instabilities in anisotropic semiconductors and is based on Refs. [100–105]. As was shown in previous parts, properties of electromagnetic waves in uniaxial conducting crystals differ considerably from those in an isotropic medium. When a static strong electric field is applied in a direction not parallel to the crystal axis, the electric field of the extraordinary wave with a wave vector $\boldsymbol{k}\perp\boldsymbol{E}_0$ is not parallel to $\boldsymbol{E}_0$ and exhibits an elliptical polarization. Moreover, in contrast to an isotropic medium, for a uniaxial semiconductor all three principal values of the differential conductivity tensor are different, and the principal axes of the tensor are not parallel to the crystal

Radiowaves and Polaritons in Anisotropic Media. Roland H. Tarkhanyan and Nikolaos K. Uzunoglu

ISBN: 3-527-40615-8

axis and to $\boldsymbol{E}_0$. In such a situation, the excitation and amplification of the waves exhibit a number of interesting features that will be studied in the present part.

In Chapter 9 it will be shown that an extraordinary wave can be excited in uniaxial semiconductors not only for a negative but also for a positive differential conductivity, that is, in the absence of a falling region in the current–voltage characteristic describing the current in the direction of $\boldsymbol{E}_0$. It will be shown also that, in contrast to isotropic materials, the anisotropy leads to the electromagnetic instability of the waves traveling along the applied field $\boldsymbol{E}_0$.

We will see that the components of the differential conductivity tensor have different signs and, in some cases, the "antidissipation" related to the contribution of terms with a negative resistivity to the alternating current can be greater than the damping. Waves in a crystal with a positive differential resistivity can be amplified provided that the conductivity is anisotropic and the crystal lattice permittivity is also anisotropic. These two conditions distinguish the instability under consideration from that in the anisotropic plasma, which occurs in the absence of a static current [106, 107]. We will calculate the growth rates of the waves which depend on the mutual orientation of the crystal axis and the field $\boldsymbol{E}_0$. Amplification of the waves takes place in a definite interval of the angles between these two directions. Excitation conditions as well as the frequencies of excited waves will be found also.

In Chapter 10, a macroscopic theory of electromagnetic instability for the surface waves is presented. It will be shown that the waves propagating along the interface between the crystal and a vacuum can be excited only if the retardation of the Coulomb interaction is taken into account. In contrast to the bulk waves, the surface waves traveling along the static electric field, which is parallel to the interface, may become unstable even for an isotropic (in its natural state) narrow-gap semiconductor. Under steady-state conditions, the energy of the waves is emitted into the vacuum. The instability criteria, as well as the frequency and the propagation direction of the emitted waves, will be determined. Nonradiative instability of the waves localized at the interface between narrow-gap and wide-gap semiconductor materials will also be considered. The final sections of Chapter 10 deal with instabilities of the quasistatic interface waves in two- and three-layered heterostructures and include a detailed description of the influence of two-dimensional mobile hot carriers on the instabilities of guided charge density waves.

9
Excitation and Amplification of the Bulk Electromagnetic Waves

9.1
Differential Conductivity Tensor

Consider a uniaxial semiconductor in the presence of a strong static electric field $\boldsymbol{E}_0$. The current density is given by

$$\boldsymbol{J}_0 = \sigma \boldsymbol{E}_0 \tag{9.1}$$

where the conductivity tensor σ has two different eigenvalues: $\sigma_{||}$ and $\sigma_{\perp}$ along and across the crystal axis, respectively. Let both $\sigma_{||}$ and $\sigma_{\perp}$ be given as functions of E_0^2. A small change (fluctuation) $\boldsymbol{E}$ in $\boldsymbol{E}_0$ causes a change

$$\boldsymbol{J} = \sigma^{\mathrm{d}} \boldsymbol{E} \tag{9.2}$$

in $\boldsymbol{J}_0$, where

$$\sigma_{ij}^{\mathrm{d}} = \frac{\partial J_{0i}}{\partial E_{0j}} = \sigma_{ij} + \frac{\partial \sigma_{il}}{\partial E_{0j}} E_{0l} \tag{9.3}$$

is the differential conductivity tensor. In the rectangular coordinate system in which $\boldsymbol{E}_0$ is parallel to the z-axis and the crystal axis lies in the xz plane and makes an angle θ with the z-axis, the conductivity tensor has a symmetric form

$$\sigma = \begin{pmatrix} \sigma_{xx} & 0 & \sigma_{xz} \\ 0 & \sigma_{\perp} & 0 \\ \sigma_{xz} & 0 & \sigma_{zz} \end{pmatrix} \tag{9.4}$$

where

$$\sigma_{xx} = \sigma_{\perp} \cos^2\theta + \sigma_{||} \sin^2\theta \tag{9.4a}$$

Radiowaves and Polaritons in Anisotropic Media. Roland H. Tarkhanyan and Nikolaos K. Uzunoglu

ISBN: 3-527-40615-8

$$\sigma_{zz} = \sigma_{\|}\cos^2\theta + \sigma_{\perp}\sin^2\theta \tag{9.4b}$$

$$\sigma_{xz} = (\sigma_{\|} - \sigma_{\perp})\sin\theta\cos\theta \tag{9.4c}$$

Using Eqs. (9.3), (9.4a), (9.4b) and (9.4c), after a simple calculation we obtain

$$\boldsymbol{\sigma}^{\mathrm{d}} = \begin{pmatrix} \sigma_{xx} & 0 & \sigma_{xz}^{\mathrm{d}} \\ 0 & \sigma_{\perp} & 0 \\ \sigma_{xz} & 0 & \sigma_{zz}^{\mathrm{d}} \end{pmatrix} \tag{9.5}$$

where

$$\sigma_{xz}^{\mathrm{d}} = (\sigma_{\|}^{\mathrm{d}} - \sigma_{\perp}^{\mathrm{d}})\sin\theta\cos\theta \tag{9.5a}$$

$$\sigma_{zz}^{\mathrm{d}} = \sigma_{\|}^{\mathrm{d}}\cos^2\theta + \sigma_{\perp}^{\mathrm{d}}\sin^2\theta \tag{9.5b}$$

and

$$\sigma_{\|,\perp}^{\mathrm{d}} = \sigma_{\|,\perp}\left(1 + 2\frac{\mathrm{d}\ln\sigma_{\|,\perp}}{\mathrm{d}\ln E_0^2}\right) \tag{9.6}$$

Note that both $\sigma_{\|}^{\mathrm{d}}$ and $\sigma_{\perp}^{\mathrm{d}}$ describe the differential conductivity along the field $\boldsymbol{E}_0$ for $\theta = 0$ and $\theta = \pi/2$, respectively. We have to note also that the tensor $\boldsymbol{\sigma}^{\mathrm{d}}$, unlike $\boldsymbol{\sigma}$, is not symmetric: $\boldsymbol{\sigma}_{ij}^{\mathrm{d}} \neq \boldsymbol{\sigma}_{ji}^{\mathrm{d}}$. It means that $\boldsymbol{\sigma}^{\mathrm{d}}$ has, in general, three different principal values, that is, it is a biaxial tensor. For example, in the case $\theta = \pi/2$, the principal values are $\boldsymbol{\sigma}_{\|}, \boldsymbol{\sigma}_{\perp}$ and $\boldsymbol{\sigma}_{\perp}^{\mathrm{d}}$. For $\theta = 0$ and for an isotropic medium, where $\boldsymbol{\sigma}_{\perp} = \boldsymbol{\sigma}_{\|} = \boldsymbol{\sigma}$, the tensor $\boldsymbol{\sigma}^{\mathrm{d}}$ is a uniaxial one with two different eigenvalues: σ and σ_{d}.

For $\theta \neq 0, \pi/2$, $\boldsymbol{\sigma}^{\mathrm{d}}$ can be decomposed into symmetric and antisymmetric parts:

$$\boldsymbol{\sigma}^{\mathrm{d}} = \boldsymbol{\sigma}_{\mathrm{s}}^{\mathrm{d}} + \boldsymbol{\sigma}_{\mathrm{a}}^{\mathrm{d}} \tag{9.7}$$

By an appropriate orthogonal transformation, using a unitary transform matrix, $\sigma_{\mathrm{s}}^{\mathrm{d}}$ can be transformed into a diagonal form:

$$\boldsymbol{\sigma}_{\mathrm{s}}^{\mathrm{d}} = \begin{pmatrix} \frac{1}{2}(\sigma_1 + \sigma_2) & 0 & 0 \\ 0 & \sigma_{\perp} & 0 \\ 0 & 0 & \frac{1}{2}(\sigma_1 - \sigma_2) \end{pmatrix} \tag{9.8}$$

where

$$\sigma_1 = (\sigma_{\perp} + \sigma_{\|}^{\mathrm{d}})\cos^2\theta + (\sigma_{\|} + \sigma_{\perp}^{\mathrm{d}})\sin^2\theta \tag{9.8a}$$

$$\sigma_2 = [(\sigma_\perp - \sigma_{||}^{d})^2 \cos^2\theta + (\sigma_{||} - \sigma_\perp^{d})^2 \sin^2\theta]^{1/2} \tag{9.8b}$$

while σ_a^d remains antisymmetric and can be written in the form

$$\boldsymbol{\sigma}_a^d = \begin{pmatrix} 0 & 0 & \sigma_3 \\ 0 & 0 & 0 \\ -\sigma_3 & 0 & 0 \end{pmatrix} \tag{9.9}$$

where

$$\sigma_3 = \frac{1}{2}[\sigma_\perp - \sigma_{||} - (\sigma_\perp^d - \sigma_{||}^d)]\sin\theta\cos\theta \tag{9.10}$$

9.2 Dispersion Relations for the Waves in the Presence of a Strong Static Electric Field E_0

Let us consider a small electromagnetic perturbation of the form $\boldsymbol{E}$, $\boldsymbol{H}$ ~$\exp[i(k_x x + k_z z - \omega t)]$, where the wave angular frequency ω is much smaller compared to the frequency of the charge-carrier collisions τ^{-1}, but is much more than the scalar product of the wave vector $\boldsymbol{k}$ and the drift velocity $\boldsymbol{v}_0$:

$$\boldsymbol{k}\boldsymbol{v}_0 \ll \omega \ll \tau^{-1} \tag{9.11}$$

In this interval of ω one may neglect the frequency dependence of the conductivity and the spatial dispersion caused by a drift current. Eliminating the magnetic field

$$\boldsymbol{H} = -i(\mu_0\omega)^{-1}\text{rot}\boldsymbol{E} \tag{9.12}$$

from Maxwell's equations linearized in small fluctuations and using Eqs. (9.2) and (9.5) for the current-density fluctuation, we find an ordinary TE-polarized wave, for which $\boldsymbol{E} = (0, E, 0)$, and an extraordinary TM-polarized wave, for which $\boldsymbol{H} = (0, H, 0)$.

The ordinary wave is described by the dispersion equation

$$\frac{c^2k^2}{\omega^2} = \varepsilon_\perp + i\frac{\sigma_\perp}{\varepsilon_0\omega} \tag{9.13}$$

where $\varepsilon_\perp$ is the crystal lattice static dielectric constant in the plane perpendicular to the crystal axis. Solving Eq. (9.13) for ω, we find that the frequency is given by

$$\omega = \pm\left(\frac{c^2k^2}{\varepsilon_\perp} - \gamma_0^2\right)^{1/2} + i\gamma_0 \tag{9.14}$$

where

$$\gamma_0 = -\frac{\sigma_\perp}{2\varepsilon_0\varepsilon_\perp} \tag{9.14a}$$

is negative. It follows that for $k > \sigma_\perp(2c\varepsilon_0\sqrt{\varepsilon_\perp})^{-1}$ the ordinary wave is damped exponentially with time $[E \sim \exp(\gamma_0 t)]$. Therefore, there is no instability.

The electric field vector for the extraordinary wave lies in the xz-plane and is elliptically polarized:

$$\frac{E_x}{E_z} = \frac{c^2 k_x k_z + \varepsilon_{xz}\omega^2 + \mathrm{i}\omega\sigma_{xz}^{\mathrm{d}}\varepsilon_0^{-1}}{c^2 k_z^2 - \varepsilon_{xx}\omega^2 - \mathrm{i}\omega\sigma_{xx}\varepsilon_0^{-1}} \tag{9.15}$$

where ε_{ij} are components of the crystal lattice static dielectric tensor given by

$$\varepsilon = \begin{pmatrix} \varepsilon_\perp\cos^2\theta + \varepsilon_{||}\sin^2\theta & 0 & (\varepsilon_{||} - \varepsilon_\perp)\sin\theta\cos\theta \\ 0 & \varepsilon_\perp & 0 \\ (\varepsilon_{||} - \varepsilon_\perp)\sin\theta\cos\theta & 0 & \varepsilon_\perp\sin^2\theta + \varepsilon_{||}\cos^2\theta \end{pmatrix} \tag{9.16}$$

The dispersion relation for the extraordinary wave can be written in the form of an equation cubic in ω:

$$A_0\omega^3 + \mathrm{i}A_1\omega^2 - (A_2 + A_3)\omega - \mathrm{i}A_4 = 0 \tag{9.17}$$

where

$$A_0 = \varepsilon_0\varepsilon_{||}\varepsilon_\perp, \qquad A_1 = (\varepsilon_{||}\sigma_\perp + \varepsilon_\perp\sigma_{||}^{\mathrm{d}})\cos^2\theta + (\varepsilon_{||}\sigma_\perp^{\mathrm{d}} + \varepsilon_\perp\sigma_{||})\sin^2\theta \tag{9.17a}$$

$$A_2 = \varepsilon_0 c^2(\varepsilon_{xx}k_x^2 + \varepsilon_{zz}k_z^2 + 2\varepsilon_{xz}k_x k_z), \qquad A_3 = \varepsilon_0^{-1}(\sigma_\perp\sigma_{||}^{\mathrm{d}}\cos^2\theta + \sigma_{||}\sigma_\perp^{\mathrm{d}}\sin^2\theta) \tag{9.17b}$$

$$A_4 = c^2[\sigma_{xx}k_x^2 + \sigma_{zz}^{\mathrm{d}}k_z^2 + (\sigma_{xz} + \sigma_{xz}^{\mathrm{d}})k_x k_z] \tag{9.17c}$$

In a frequency region for which

$$\mathrm{Re}\omega >> \nu, \qquad \nu \equiv \max\left\{\frac{\sigma_\perp}{\varepsilon_0\varepsilon_\perp}, \frac{\sigma_{||}}{\varepsilon_0\varepsilon_{||}}, \frac{\sigma_\perp^{\mathrm{d}}}{\varepsilon_0\varepsilon_\perp}, \frac{\sigma_{||}^{\mathrm{d}}}{\varepsilon_0\varepsilon_{||}}\right\} \tag{9.18}$$

Eq. (9.17) has an approximate solution

$$\omega \approx \pm\sqrt{\frac{A_2}{A_0}} + \mathrm{i}\gamma \tag{9.19}$$

$$\gamma = \frac{1}{2}\left(\frac{A_4}{A_2} - \frac{A_1}{A_0}\right) \tag{9.20}$$

where $\mathrm{Re}\omega >> \gamma$. It is evident that if γ is positive, the wave amplitude increases exponentially with time, that is the wave is unstable.

Let us consider now some simple cases.

a. The static electric field $\boldsymbol{E}_0$ is parallel to the crystal axis $\boldsymbol{C}$.
In this case $\theta = 0$ and from Eq. (9.20) we obtain

$$\gamma = -\frac{\sigma_\perp + \beta^2\sigma_{||}^{\mathrm{d}}\tan^2\varphi}{2\varepsilon_0\varepsilon_\perp(1+\beta\tan^2\varphi)} \tag{9.21}$$

where φ is the angle between the wave vector $\boldsymbol{k}$ and $\boldsymbol{E}_0$ and

$$\beta \equiv \varepsilon_\perp/\varepsilon_{||} \tag{9.22}$$

Using Eq. (9.21), we conclude that an electromagnetic instability occurs only if the following conditions are fulfilled:

$$\sigma_{||}^{\mathrm{d}} < 0 \qquad |\sigma_{||}^{\mathrm{d}}| > \sigma_\perp\beta^{-2}\cot^2\varphi \tag{9.23}$$

For a given negative value of $\sigma_{||}^{\mathrm{d}}$ the instability is only possible for the waves for which the angle φ is restricted to the range $\varphi > \varphi_{||}$, where

$$\cot\varphi_{||} = \beta\sqrt{|\sigma_{||}^{\mathrm{d}}|\sigma_\perp^{-1}} \tag{9.24}$$

b. $\boldsymbol{E}_0 \perp \boldsymbol{C}$ $(\theta = \pi/2)$.
In this case the instability condition is expressed in the form

$$\sigma_\perp^{\mathrm{d}} < 0 \qquad |\sigma_\perp^{\mathrm{d}}| > \sigma_{||}\beta^2\cot^2\varphi \tag{9.25}$$

It follows that at a given negative value of $\sigma_\perp^{\mathrm{d}}$ the waves can only be amplified for which $\varphi > \varphi_\perp$, where

$$\tan\varphi_\perp = \beta\sqrt{\sigma_{||}|\sigma_\perp^{\mathrm{d}}|^{-1}} \tag{9.26}$$

c. In the case of optically isotropic (in their natural state) semiconductors, the imaginary part of ω is given by

$$\gamma = -(\sigma\cos^2\varphi + \sigma_{\mathrm{d}}\sin^2\varphi)(2\varepsilon_0\varepsilon)^{-1} \tag{9.27}$$

from which we obtain the instability condition

$$\sigma_{\mathrm{d}} < 0, \qquad |\sigma_{\mathrm{d}}| > \sigma \cot^2 \varphi \tag{9.28}$$

Therefore, in the voltage range in which NDR occurs, the waves for which $\tan\varphi < \sqrt{\sigma|\sigma_{\mathrm{d}}|^{-1}}$ (in particular, the waves propagating along $\boldsymbol{E}_0$) are stable. We have to note that in a uniaxial semiconductor, in contrast to the isotropic one, the extraordinary wave with $\boldsymbol{k}||\boldsymbol{E}_0$ can be unstable for negative as well as for positive values of both $\sigma_{\perp}^{\mathrm{d}}$ and $\sigma_{||}^{\mathrm{d}}$ (see Section 9.5).

9.3
Instability of the Waves with $\boldsymbol{k} \perp \boldsymbol{E}_0$

In this section we will discuss extraordinary waves propagating along the direction perpendicular to the electric field $\boldsymbol{E}_0$, for an arbitrary value of the angle θ between $\boldsymbol{E}_0$ and the crystal axis. Consider a uniaxial semiconductor infinite in the x-direction and let us analyze growth rates for the waves traveling along the x-axis. Using Eqs. (9.19) and (9.20), we find the real and imaginary parts of ω in the form

$$\mathrm{Re}\,\omega = \pm ck\left(\frac{\cos^2\theta}{\varepsilon_{||}} + \frac{\sin^2\theta}{\varepsilon_{\perp}}\right)^{1/2} \tag{9.29}$$

$$\gamma = -\frac{1}{2\varepsilon_0\varepsilon_{\perp}}\left[\sigma_{\perp}^{\mathrm{d}}\sin^2\theta + \beta\sigma_{||}^{\mathrm{d}}\cos^2\theta + \frac{\rho\sin^2\theta\cos^2\theta}{\cos^2\theta + \beta^{-1}\sin^2\theta}\right] \tag{9.30}$$

where

$$\rho \equiv (\beta - 1)(\sigma_{||} - \sigma_{\perp}\beta^{-1}) \tag{9.31}$$

and β is given by Eq. (9.22).

Now let us consider the following possibilities [100]:

a. $\sigma_{\perp}^{\mathrm{d}} < 0$ and $\sigma_{||}^{\mathrm{d}} < 0$.

In this case, the function $\gamma(\theta)$ depends on the sign of the parameter ς and on the ratio

$$a = \frac{\varsigma^2}{4\sigma_{||}^{\mathrm{d}}\sigma_{\perp}^{\mathrm{d}}} \tag{9.32}$$

where

$$\varsigma = \rho + \sigma_{||}^{\mathrm{d}} + \sigma_{\perp}^{\mathrm{d}} \tag{9.33}$$

For $\varsigma > 0$ and $a > 1$, the wave is amplified only in two intervals of the angle $\theta : \theta < \theta_-$ and $\theta > \theta_+$, where $\theta_\pm$ are given by solutions of the equation $\gamma(\theta) = 0$:

$$\tan^2\theta_{\pm} = \frac{\beta}{2}\left[-\frac{\varsigma}{\sigma_{\perp}^{d}} \pm \sqrt{\left(\frac{\varsigma}{\sigma_{\perp}^{d}}\right)^{2} - \frac{4\sigma_{\parallel}^{d}}{\sigma_{\perp}^{d}}}\right] \tag{9.34}$$

The growth rate is a monotonically decreasing function of the angle θ in the region $\theta < \theta_-$ and it is an increasing function in the region $\theta > \theta_+$ (see Fig. 9.1a, continuous curves). Dashed curves in this figure correspond to the case $a = 1$; then

$$\tan\theta_+ = \tan\theta_- = \beta^{1/2}\left(\frac{\sigma_{\parallel}^{d}}{\sigma_{\perp}^{d}}\right)^{1/4} \tag{9.35}$$

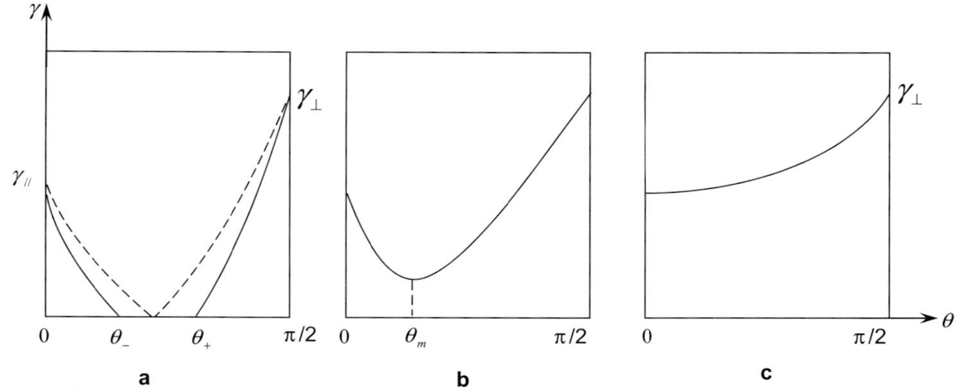

Fig. 9.1 Angle dependence of the growth rate for the case $\sigma_{\parallel,\perp}^{d} < 0$. It is assumed that $\gamma_{\perp} > \gamma_{\parallel}$. a. $\varsigma > 0, a \geq 1$; b. $\varsigma > 0, a < 1$; c. $\varsigma < 0$.

For $\varsigma > 0$ and $a < 1$, the wave is amplified for arbitrary θ. The function $\gamma(\theta)$ decreases initially from the value

$$\gamma_{\parallel} = (2\varepsilon_0\varepsilon_{\parallel})^{-1}|\sigma_{\parallel}^{d}| \tag{9.36}$$

at $\theta = 0$, reaches a minimum at $\theta = \theta_{\mathrm{m}}$ given by

$$\gamma_{\mathrm{m}} = \gamma_{\parallel}\cos^2\theta_{\mathrm{m}}\left[1 + \frac{\tan^2\theta_{\mathrm{m}}}{|\sigma_{\parallel}^{d}|}\left(\frac{|\sigma_{\perp}^{d}|}{\beta} - \frac{\rho}{\beta + \tan^2\theta_{\mathrm{m}}}\right)\right] \tag{9.37}$$

$$\tan^2\theta_{\mathrm{m}} = \left(\varsigma_0 + \sqrt{\varsigma_0^2 + \beta\varsigma_1\varsigma_2}\right)\varsigma_1^{-1} \tag{9.38}$$

$$\varsigma_0 = \sigma_{\perp}^{d} - \beta\sigma_{\|}^{d} \qquad \varsigma_1 = \varsigma + |\sigma_{\perp}^{d}|(1+\beta^{-1}) \qquad \varsigma_2 = \varsigma + |\sigma_{\|}^{d}|(1+\beta) \tag{9.38a}$$

and then increases monotonically to the value

$$\gamma_{\perp} = (2\varepsilon_0\varepsilon_{\perp})^{-1}|\sigma_{\perp}^{d}| \tag{9.39}$$

at $\theta = \pi/2$ (see Fig. 9.1b).

For $\varsigma < 0$, the wave is again amplified for arbitrary θ, and γ is a monotonically increasing function of θ (see Fig. 9.1c).

b. $\sigma_{\perp}^{d} > 0$ and $\sigma_{\|}^{d} < 0$.

In this case, an amplification of the waves can only occur in the region $\theta > \theta_+$; γ is either a monotonically decreasing function of θ (see Fig. 9.2a, dashed curve for $\varsigma > 0$) or it increases initially, reaches a maximum at $\theta = \theta_M$ given by

$$\gamma_M = \gamma_{\|}\cos^2\theta_M\left[1 + \frac{\tan^2\theta_M}{\sigma_{\|}^{d}}\left(\frac{\sigma_{\perp}^{d}}{\beta} + \frac{\rho}{\beta + \tan^2\theta_M}\right)\right] \tag{9.40}$$

$$\tan^2\theta_M = \frac{\beta\sigma_{\|}^{d} - \sigma_{\perp}^{d} + \sqrt{\rho[\beta\rho - (1-\beta)(\sigma_{\perp}^{d} - \beta\sigma_{\|}^{d})}}{|\varsigma| + \sigma_{\perp}^{d}(1+\beta^{-1})} \tag{9.41}$$

and then decreases to zero at $\theta = \theta_+$ (Fig. 9.2a, continuous curve). The latter case occurs provided that the following condition is satisfied:

$$\varsigma < 0 \quad |\varsigma| > (1+\beta)|\sigma_{\|}^{d}| \tag{9.42a}$$

which is equivalent to

$$\rho < 0 \quad |\rho| > \sigma_{\perp}^{d} + \beta|\sigma_{\|}^{d}| \tag{9.42b}$$

The condition $\rho < 0$ can only be satisfied for materials for which the following inequalities hold:

$$\frac{\sigma_{\perp}}{\sigma_{\|}} < \frac{\varepsilon_{\perp}}{\varepsilon_{\|}} < 1, \quad \text{or} \quad \frac{\sigma_{\perp}}{\sigma_{\|}} > \frac{\varepsilon_{\perp}}{\varepsilon_{\|}} > 1 \tag{9.43}$$

c. $\sigma_{\perp}^{d} < 0$ and $\sigma_{\|}^{d} > 0$. In this case, the wave is amplified only in the region $\theta > \theta_+$; γ increases monotonically as a function of θ from zero at $\theta = \theta_+$ until it reaches the value $\gamma_{\perp}$ at $\theta = \pi/2$ (Fig. 9.2b).

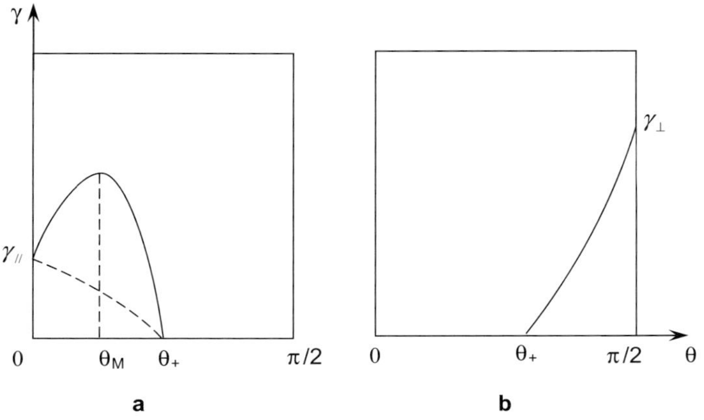

Fig. 9.2 The function $\gamma(\theta)$ for the cases $\sigma_\perp^d > 0, \sigma_\parallel^d < 0$ (a) and $\sigma_\perp^d < 0, \sigma_\parallel^d > 0$ (b).

9.4 Effective Differential Conductivity. Instability in the Absence of a Falling Region in the Current–Voltage Characteristic

Consider now a sample for which both $\sigma_\parallel^d$ and $\sigma_\perp^d$ are positive. This case is of special interest since the amplification of the waves with $\boldsymbol{k}\perp\boldsymbol{E}_0$ takes place in the absence of a falling region in the current–voltage characteristic, which is impossible for an isotropic material. We have to note that the physical mechanism governing the growth of the extraordinary wave in this case is the same as in the case of negative differential conductivity: the conduction electrons interacting with an alternating electric field lose part of their energy to the field and, therefore, the field amplitude grows and the system under study becomes unstable. We can explain qualitatively this effect as follows [100]. A weak alternating field $\boldsymbol{E}$ parallel to the xz-plane gives rise to a weak alternating current which is not parallel to $\boldsymbol{E}$, in contrast to the isotropic case. The magnetic field of the wave that is parallel to the y-axis bends the electron trajectories and results in a Hall component of the alternating current in the xz-plane. The angle ψ between the current density $\boldsymbol{J}$ and the field vector $\boldsymbol{E}$ can be greater than $\pi/2$ and, therefore, $\boldsymbol{JE} = JE\cos\psi$ is negative. Consequently, the energy dissipation of the wave is negative and the wave gains energy from charge carriers and the corresponding energy loss is compensated by the source creating a static field.

Using Eq. (9.2), for $k_z = 0$ we obtain $\boldsymbol{JE} = \sigma_{\text{ef}}^d(\theta)E^2$, where

$$\sigma_{\text{ef}}^d(\theta) = (1+\xi^2)^{-1}[\sigma_{xx} + \xi^2\sigma_{zz}^d + \xi(\sigma_{xz} + \sigma_{xz}^d)] \tag{9.44}$$

$$\xi \equiv \frac{E_z}{E_x} = -\frac{\varepsilon_{xx} + \mathrm{i}(\varepsilon_0\omega)^{-1}\sigma_{xx}}{\varepsilon_{xz} + \mathrm{i}(\varepsilon_0\omega)^{-1}\sigma_{xz}^d} \tag{9.45}$$

Therefore, amplification or damping of the extraordinary wave in a uniaxial crystal is governed by the sign of the effective differential conductivity given by Eq. (9.44) rather than by the signs of $\sigma_{\|}^{\mathrm{d}}$ and $\sigma_{\perp}^{\mathrm{d}}$, which appear in the expression for $\sigma_{\mathrm{ef}}^{\mathrm{d}}$ as parameters. In the frequency region defined by Eq. (9.18), we may put $\xi \cong -\varepsilon_{xx}/\varepsilon_{xz}$ and then the growth rate given by Eq. (9.30) is related to $\sigma_{\mathrm{ef}}^{\mathrm{d}}$ by the following simple expression:

$$\gamma = -\frac{\sigma_{\mathrm{ef}}^{\mathrm{d}}(\theta)(\beta^2 + \tan^2\theta)}{2\varepsilon_0\varepsilon_{\perp}(\beta + \tan^2\theta)} \tag{9.46}$$

For an isotropic medium ($\beta = 1$, $\sigma_{\perp,\|}^{\mathrm{d}} = \sigma_{\mathrm{d}}$) we obtain $\sigma_{\mathrm{ef}}^{\mathrm{d}} = \sigma_{\mathrm{d}}$ and the wave is amplified provided that $\sigma_{\mathrm{d}} < 0$. For a uniaxial crystal satisfying the conditions $\sigma_{\|}^{\mathrm{d}} > 0$ and $\sigma_{\perp}^{\mathrm{d}} > 0$, we find that $\sigma_{\mathrm{ef}}^{\mathrm{d}}(\theta)$ is negative provided that

$$\rho < -\left(\sqrt{\sigma_{\perp}^{\mathrm{d}}} + \sqrt{\sigma_{\|}^{\mathrm{d}}}\right)^2 \tag{9.47}$$

and

$$\theta_- < \theta < \theta_+ \tag{9.48}$$

where ρ and $\theta_{\pm}$ are given by Eqs. (9.31) and (9.34), respectively.

We have to note again that the condition (9.47) can only be satisfied for materials with anisotropic conductivity and with anisotropic permittivity of the crystal lattice, for which the inequalities (9.43) hold.

When the angle θ lies in the interval defined by Eq. (9.48), the growth rate increases initially, reaches a maximum at $\theta = \theta_{\mathrm{M}}$ [see Eq. (9.41)] and then decreases and vanishes at $\theta = \theta_+$ (Fig. 9.3). The maximum value of γ is given by Eq. (9.40) and can also be written in an equivalent form

$$\gamma_{\mathrm{M}} = \frac{\sigma_{\perp}^{\mathrm{d}}(\tan^2\theta_{\mathrm{M}} - \tan^2\theta_-)(\tan^2\theta_+ - \tan^2\theta_{\mathrm{M}})}{2\varepsilon_0\varepsilon_{\perp}(1 + \tan^2\theta_{\mathrm{M}})(\beta + \tan^2\theta_{\mathrm{M}})} \tag{9.49}$$

9.5
Instability of the Waves Propagating along $\boldsymbol{E}_0$

As we have seen above (see Section 9.2), the waves propagating along the static field $\boldsymbol{E}_0$ in an isotropic material cannot be amplified even when the differential conductivity is negative. In contrast to this, now we will see that the anisotropy of a semiconductor leads to the instability of the extraordinary wave traveling along $\boldsymbol{E}_0$. Moreover, such an instability can occur not only for negative but also for positive values of the differential conductivity σ_{zz}^{d} along $\boldsymbol{E}_0$.

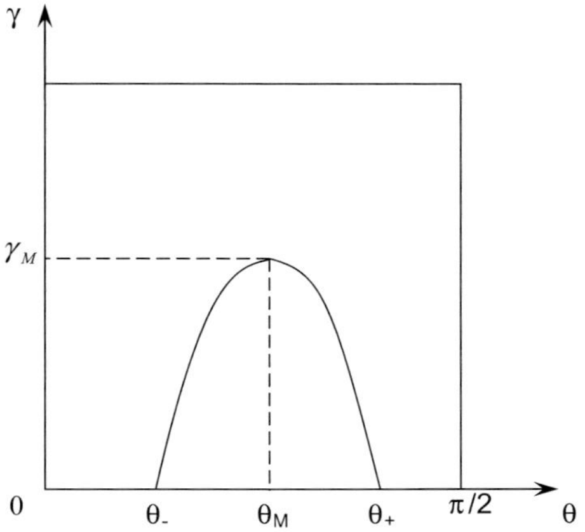

Fig. 9.3 Angle dependence of the growth rate for positive values of both $\sigma_{||}^{\mathrm{d}}$ and $\sigma_{\perp}^{\mathrm{d}}$.

Substituting $k_x = 0$ and $k_z = k$ in Eqs. (9.19) and (9.20) and using (9.17a)–(9.17c), for the frequency and growth rate of the waves with $\boldsymbol{k}||\boldsymbol{E}_0$ we obtain [101]

$$\mathrm{Re}\,\omega = \pm ck\sqrt{\frac{\cos^2\theta}{\varepsilon_\perp} + \frac{\sin^2\theta}{\varepsilon_{||}}} \tag{9.50}$$

$$\gamma = -\frac{1}{2\varepsilon_0\varepsilon_\perp}\left[\sigma_\perp\cos^2\theta + \beta\sigma_{||}\sin^2\theta + \frac{\rho_\mathrm{d}\sin^2\theta\cos^2\theta}{\cos^2\theta + \beta\sin^2\theta}\right] \tag{9.51}$$

where

$$\rho_\mathrm{d} = (1-\beta)(\sigma_\perp^\mathrm{d} - \beta\sigma_{||}^\mathrm{d}) \tag{9.51a}$$

After a simple calculation the instability condition $\gamma > 0$ may be written as

$$\rho_\mathrm{d} < -\beta\left(\sqrt{\sigma_\perp} + \sqrt{\sigma_{||}}\right)^2 \tag{9.52}$$

$$\tilde{\theta}_- < \theta < \tilde{\theta}_+ \tag{9.53}$$

where $\tilde{\theta}_\pm$ are given by

$$\tan^2\tilde{\theta}_\pm = (2\beta\sigma_{||})^{-1}[-\tilde{\varsigma}\pm\sqrt{\tilde{\varsigma}^2 - 4\sigma_{||}\sigma_\perp}] \tag{9.54}$$

$$\tilde{\varsigma} = \sigma_{||} + \sigma_\perp + \rho_\mathrm{d}\beta^{-1} \tag{9.55}$$

Note that $\tilde{\varsigma} < 0$ when the condition (9.52) is satisfied. A plot of γ versus θ in the unstable region is similar to that shown in Fig. 9.3, where $\theta_\pm$ and θ_M should be replaced by $\tilde{\theta}_\pm$ and $\tilde{\theta}_M$, respectively. The angle $\tilde{\theta}_M$ is defined by

$$\tan^2 \tilde{\theta}_M = \frac{\sigma_\perp - \beta\sigma_{||} + \sqrt{|\rho_d|(\beta\sigma_{||} - \tilde{\varsigma} + \beta^{-1}\sigma_\perp)}}{|\rho_d| + \beta(\beta\sigma_{||} - \sigma_\perp)} \tag{9.56}$$

The maximum value of γ at $\theta = \tilde{\theta}_M$ is given by

$$\tilde{\gamma}_M = \frac{\sigma_{||}(\tan^2 \tilde{\theta}_M - \tan^2 \tilde{\theta}_-)(\tan^2 \tilde{\theta}_+ - \tan^2 \tilde{\theta}_M)}{2\varepsilon_0\varepsilon_{||}(1 + \tan^2 \tilde{\theta}_M)(\tan^2 \tilde{\theta}_M + \beta^{-1})} \tag{9.57}$$

If $\sigma_\perp^d > 0$ and $\sigma_{||}^d < 0$, the instability condition (9.52) can only be fulfilled in crystals for which $\varepsilon_\perp > \varepsilon_{||}$, and *vice versa*. If both $\sigma_\perp^d$ and $\sigma_{||}^d$ are negative, the necessary condition is

$$\beta > \max\{1, \sigma_\perp^d/\sigma_{||}^d\} \qquad \text{or} \qquad \beta < \min\{1, \sigma_\perp^d/\sigma_{||}^d\} \tag{9.58}$$

It is noticeable that the waves can also be unstable when both $\sigma_\perp^d$ and $\sigma_{||}^d$ are positive. In this case it is necessary that the condition

$$\frac{\sigma_\perp^d}{\sigma_{||}^d} > \frac{\varepsilon_\perp}{\varepsilon_{||}} > 1, \qquad \text{or} \qquad \frac{\sigma_\perp^d}{\sigma_{||}^d} < \frac{\varepsilon_\perp}{\varepsilon_{||}} < 1 \tag{9.59}$$

is satisfied.

Now we wish to show that when $\sigma_{\perp,||}^d > 0$, the wave instability occurs in the absence of a potential electric field domain. To see this, let us consider the dispersion relation for potential (longitudinal) waves in the absence of retardation effects. In this limiting case ($c \to \infty$), Eq. (9.17) gives

$$\omega = -iA_4A_2^{-1} \tag{9.60}$$

Note that for the waves with $\boldsymbol{k} \perp \boldsymbol{E}_0$, Eq. (9.60) leads to $\omega = -i\sigma_{xx}(\varepsilon_0\varepsilon_{xx})^{-1}$; it follows that the longitudinal wave is damped exponentially with time and, therefore, there is no potential electric field domain. For the waves with $\boldsymbol{k} || \boldsymbol{E}_0$, Eq. (9.60) gives

$$\omega = -i\sigma_{zz}^d(\varepsilon_0\varepsilon_{zz})^{-1} \tag{9.61}$$

This result means that the potential wave propagating along $\boldsymbol{E}_0$ can be amplified provided that σ_{zz}^d is negative:

$$\sigma_{||}^d \cos^2\theta + \sigma_\perp^d \sin^2\theta < 0 \tag{9.62}$$

It follows that for $\sigma_{\perp}^{d} > 0$ and $\sigma_{\|}^{d} > 0$, amplification of the potential wave is impossible and, therefore, electromagnetic instability occurs in the absence of a domain. For $\sigma_{zz}^{d} < 0$, the analogous situation is only possible if the domain-formation time $\sim|\omega|^{-1}$ is greater compared to the charge-carrier time of flight through the sample, which is proportional to L_z/v_{0z} (L_z is the sample length and v_{0z} is the drift velocity along $\boldsymbol{E}_0$). Using Eq. (9.61) and the relation

$$v_{0z} = \frac{J_{0z}}{qN_0} = \frac{\sigma_{zz}E_0}{qN_0}$$

where q and N_0 are the carrier charge and density, we obtain the corresponding condition in the form

$$N_0 L_z < \frac{\varepsilon_0 \varepsilon_{zz} \sigma_{zz} E_0}{q|\sigma_{zz}^{d}|} \tag{9.64}$$

Thus, when the condition (9.64) is fulfilled, the space charge decreases and the extraordinary electromagnetic wave can be amplified in the absence of the electric field domain.

Now let us go back to the instability condition given by Eq. (9.52) and note that for an isotropic material ($\beta = 1$) it cannot be satisfied. The physical explanation is simple: for an isotropic material the electric field $\boldsymbol{E}$ in the wave with $\boldsymbol{k}||\boldsymbol{E}_0$ is perpendicular to the static field $\boldsymbol{E}_0$ and the current density $\boldsymbol{J} = \sigma\boldsymbol{E}$. Then, we have $\boldsymbol{JE} = \sigma E^2 > 0$; this means that the wave is damped, and there is no instability. Unlike this, in a uniaxial semiconductor the vector $\boldsymbol{J}$ is not parallel to $\boldsymbol{E}$, and for the extraordinary wave with $\boldsymbol{k}||\boldsymbol{E}_0$ we obtain $\boldsymbol{JE} = \sigma_{ef}^{d} E^2$, where

$$\sigma_{ef}^{d} = -\frac{2\gamma\varepsilon_0\varepsilon_{\perp}(1 + \beta\tan^2\theta)}{1 + \beta^2\tan^2\theta} \tag{9.65}$$

When $\gamma > 0$, the effective differential conductivity is negative, and then the electromagnetic instability arises.

9.6 Excitation of Extraordinary Waves in a Uniaxial Semiconductor Plate

In previous sections we have investigated the wave instabilities in an infinite anisotropic medium. Now let us study the excitation of bulk electromagnetic waves in a spatially restricted plate with dimensions $L_x = d$, L_y, L_z, where $d << L_y, L_z$. We assume that the origin of the rectangular system of coordinates is situated in the center of the plate, and continue the assumption that the static field is applied in the z-direction while the crystal axis lies in the xz-plane at an angle θ to $\boldsymbol{E}_0$, as is shown in Fig. 9.4. We will restrict ourselves by consideration of the waves traveling normally to the boundaries of the plate, in the x-direction. We assume also that the plate is bounded by two insulators, whose refractive indices are n_1 in the region $x > d/2$ and n_2 in the region $x < -d/2$.

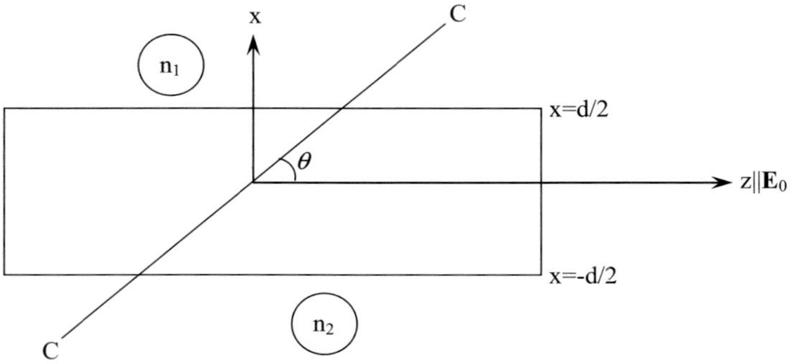

Fig. 9.4 Geometry of the uniaxial semiconductor plate. C is the optical axis of the crystal. The space regions $x > d/2$ and $x < -d/2$ are filled by isotropic insulators with refractive indices n_1 and n_2, respectively.

Using Eqs. (9.19) and (9.29), for real values of ω and complex values of k we obtain

$$ck = (\omega - i\gamma) n_0(\theta) \tag{9.66}$$

where

$$n_0(\theta) = \left(\frac{\cos^2\theta}{\varepsilon_{\|}} + \frac{\sin^2\theta}{\varepsilon_{\perp}}\right)^{-1/2} \tag{9.67}$$

and γ is given by Eq. (9.30). When the fields excited into the plate encounter another medium, some energy is reflected and some is transmitted. Taking into account that for a wave propagating in the x-direction

$$H_y = i\frac{c}{\omega}\sqrt{\frac{\varepsilon_0}{\mu_0}}\frac{dE_z}{dx} \tag{9.68}$$

standard boundary conditions for the tangential components E_z and H_y at $x = \pm d/2$ can be written in the form

$$g_+ E_+ + g_- E_- = g_1 E_1 \qquad n(g_+ E_+ - g_- E_-) = n_1 g_1 E_1 \tag{9.69a}$$

$$g_- E_+ + g_+ E_- = g_2 E_2 \qquad n(g_- E_+ - g_+ E_-) = -n_2 g_2 E_2 \tag{9.69b}$$

where

$$g_{\pm} = \exp(\pm ikd/2) \qquad g_{1,2} = \exp(ik_{1,2}d/2) \qquad k_{1,2} = n_{1,2}\omega/c \tag{9.69c}$$

Here $E_{\pm}$ are the amplitudes of the electric field for the waves traveling in opposite directions $\pm x$ into the plate and $E_{1,2}$ are those for the waves leaving the plate and propagating in the regions $x > d/2$ and $x < -d/2$, respectively, and

$$n = n_0\left(1 - i\frac{\gamma}{\omega}\right) \tag{9.70}$$

Equating the determinant of the system (9.69a) and (9.69b) to zero, we obtain the dispersion relation

$$e^{2ikd} = \frac{(n+n_1)(n+n_2)}{(n-n_1)(n-n_2)} \tag{9.71}$$

Substituting Eqs. (9.66) and (9.70) into Eq. (9.71) and separating the real and imaginary parts, we obtain the following expressions for the critical value of γ and for the frequencies of the excited waves:

$$\gamma_{\text{cr}} = \frac{c}{2n_0 d}\ln R, \tag{9.72}$$

$$\omega \equiv \omega_l = \frac{c}{n_0 d}(\pi l + \psi) \qquad l = 0, 1, 2, \ldots \tag{9.73}$$

Here R and 2ψ denote the modulus and the argument of the right-hand side of Eq. (9.71). For $\gamma_{\text{cr}}/\omega << 1$ those can be written as

$$R \cong \frac{(n_0+n_1)(n_0+n_2)}{|n_0-n_1||n_0-n_2|} \tag{9.74}$$

$$\psi \cong \frac{\gamma_{\text{cr}} n_0(n_1+n_2)(n_0^2-n_1n_2)}{\omega_l(n_0^2-n_1^2)(n_0^2-n_2^2)} \tag{9.75}$$

Excitation of the waves with discrete frequencies ω_l is only possible if the condition

$$\gamma \geq \gamma_{\text{cr}} \tag{9.76}$$

is fulfilled. Note that Eqs. (9.72)–(9.75) are only applicable in the frequency region

$$\nu << \omega_l << \tau^{-1} \tag{9.77}$$

[see Eqs. (9.11) and (9.18)] and only if

$$v_{0x} = \frac{\sigma_{xz}}{qN_0} E_0 << \frac{c}{n_0} \tag{9.78}$$

Expanding R and ψ in powers of $n_0/n_{1,2}$, in the limit $n_0 << n_{1,2}$ we obtain

$$\gamma_{cr} \cong \frac{c}{d}\left(\frac{1}{n_1} + \frac{1}{n_2}\right) \tag{9.79}$$

$$\omega_l \cong \frac{\pi c}{n_0 d} l \qquad l = 1, 2, 3, \ldots \tag{9.80}$$

Let us set $n_0 = 10$, $n_{1,2} = 10^2$, $\tau = 10^{-12}$s and $v = 10^8$s^{-1}. For $d = 1$ cm, we find that the waves with discrete frequencies $\omega_l = 10^{10} l$ (s^{-1}) can be excited.

In another limiting case $n_1 = 1$ (vacuum), $n_2 >> n_0$, we find that

$$\gamma_{cr} = \frac{c}{d}\left(\frac{1}{n_2} + \frac{1}{2n_0} \ln \frac{n_0 + 1}{|n_0 - 1|}\right) \tag{9.81}$$

and the frequencies are again given by Eq. (9.80).

Finally, in the symmetric case $n_1 = n_2 \equiv \tilde{n}$ both the even $[H_y \sim \cos(kx)]$ and the odd $[H_y \sim \sin(kx)]$ modes are excited. In this case, Eq. (9.71) gives

$$e^{ikd} = \pm \frac{k + \omega \tilde{n}/c}{k - \omega \tilde{n}/c} \tag{9.82}$$

where the upper sign corresponds to the even and the lower sign to the odd modes. Excitation conditions for the waves of different parities are identical and for $\gamma/\omega << |n_0 - \tilde{n}|/n_0$ are given by

$$\gamma > \gamma_{cr} = \frac{c}{n_0 d} \ln\left(\frac{n_0 + \tilde{n}}{|n_0 - \tilde{n}|}\right) \tag{9.83}$$

The values of the frequency are determined by Eq. (9.80), where $l = 2m$ for the even modes and $l = 2m + 1$ for the odd modes, $m = 0, 1, 2, \ldots$.

It should be noted that an excitation of the waves in a bounded crystal can occur only in a narrower interval of values of the angle θ than those corresponding to an infinite medium. For example, in the case $\sigma^{\mathrm{d}}_{\perp,\parallel} > 0$ we find that $\psi_- < \theta < \psi_+$, where tan $\psi_\pm$ are given by the solutions of the biquadratic equation $\gamma(\theta) = \gamma_{cr}$, and also the following additional condition has to be imposed: $\gamma_M > \gamma_{cr}$, where γ_M is given by Eq. (9.49). It is not difficult to see that $\psi_- > \theta_-$ and $\psi_+ < \theta_+$, where $\theta_\pm$ are given by Eq. (9.34).

9.7 Wave Amplification at Transmission through the Plate

Let us now go back to Fig. 9.4 and discuss the amplification of the electromagnetic waves incident normally on the surface of a uniaxial semiconductor slab from an isotropic insulator with a refractive index n_2. It is clear that the problem of wave amplification is of considerable importance from the point of view of possible device applications. In this respect, it is essential to find the conditions under which the wave is amplified when it leaves the plate and enters an insulator with a refractive index $n_1 = ck_1/\omega$.

We assume that the magnetic field vector in the incident wave is parallel to the y-axis; then an extraordinary refracted wave is formed in the space region $|x| < d/2$ occupied by the semiconductor slab.

Taking into account reflection at the slab boundaries, the following continuity conditions for the tangential components of the magnetic field vector can be used:

$$g_2^* H_\mathrm{i} + g_2 H_\mathrm{R} = g_- H_+ + g_+ H_- \qquad n_2^{-1}(g_2^* H_\mathrm{i} - g_2 H_\mathrm{R}) = n^{-1}(g_- H_+ - g_+ H_-) \tag{9.84a}$$

$$g_+ H_+ + g_- H_- = g_1 H_\mathrm{t} \qquad n^{-1}(g_+ H_+ - g_- H_-) = n_1^{-1} g_1 H_\mathrm{t} \tag{9.84b}$$

where H_i, H_R and H_t correspond to the amplitudes of the incident, reflected and transmitted waves, respectively, n is given by Eq. (9.70) and g_i is given by Eq. (9.69c). Eliminating H_R and $H_\pm$ from Eqs. (9.84a) and (9.84b), we obtain the following expression for the gain $\Gamma \equiv |H_\mathrm{t}/H_\mathrm{i}|$:

$$\Gamma = \frac{4n_1|n|\mathrm{e}^{a\gamma/2}}{|n - n_1||n - n_2|[R^2 - 2\mathrm{e}^{a\gamma} R\cos(a\omega - 2\psi) + \mathrm{e}^{2a\gamma}]^{1/2}} \tag{9.85}$$

where

$$a \equiv 2n_0(\theta)d/c \tag{9.86}$$

Here R and ψ are those defined in the previous section. For $\omega = \omega_l$, but $\gamma < \gamma_{\mathrm{cr}}$, we obtain $a\omega - 2\psi = 2\pi l$ and then Eq. (9.85) gives

$$\Gamma = 4n_1|n||(n - n_1)(n - n_2)\mathrm{e}^{a\gamma/2} - (n + n_1)(n + n_2)\mathrm{e}^{-a\gamma/2}|^{-1} \tag{9.87}$$

Setting $|n| \approx n_0$ and $a\gamma << 1$, from Eq. (9.87) we obtain

$$\Gamma \approx \frac{2n_1}{(n_1 + n_2)}\left[1 + \frac{\gamma d}{c}\left(\frac{n_0^2 + n_1 n_2}{n_1 + n_2}\right)\right] \tag{9.88}$$

For the symmetric case $n_1 = n_2$, Eq. (9.88) gives

$$\Gamma = 1 + \frac{\gamma d(n_0^2 + n_1^2)}{2cn_1} \tag{9.89}$$

It follows that the transmitted wave is amplified if $\gamma > 0$. For $\gamma = \gamma_{cr}$ [$R = \exp(a\gamma)$], but $\omega = \omega_l + \Delta\omega$, we find from Eq. (9.85) that

$$\Gamma = \frac{2n_1|n|\exp(-a\gamma_{cr}/2)}{|n - n_1||n - n_2|\sin(a\Delta\omega/2)} \tag{9.90}$$

which yields in the limit $a\Delta\omega << 1$, $\gamma_{cr} \sim \Delta\omega$, the following result:

$$\mathrm{H} \cong \frac{2n_1 c}{|n_0 - n_1||n_0 - n_2|d\Delta\omega} \tag{9.91}$$

Let us now make an estimate of the effect under consideration. Setting, for example, $n_0 = 10$, $n_1 = n_2 = 11$, $d = 1$ cm, $\omega = 10^{10}\mathrm{s}^{-1}$, and $\Delta\omega/\omega = 3 \times 10^{-3}$, we obtain from Eq. (9.91): $\Gamma = 2.2 \times 10^4$.

10
Instabilities of Surface Electromagnetic Waves and Excitation of Guided Charge Density Waves in Semiconductor Heterostructures

10.1
Dispersion Relation for Surface Waves in Semiconductors with Hot Bulk Carriers

In this chapter we shall develop a phenomenological theory of instability for surface electromagnetic and interface quasistatic waves whose localization depth is much greater than the crystal lattice constant and whose frequencies satisfy the condition $\omega\tau << 1$, where τ is the energy relaxation time of charge carriers.

We shall consider an isotropic (in its natural state) semiconductor (for example, n-GaAs) in the presence of a heating static electric field $\boldsymbol{E}_0$ which is parallel to the interface between the semiconductor and a vacuum or another semiconducting material. When the static field is applied, the semiconductor crystal becomes a uniaxially anisotropic one with the axis along the field.

In this section we shall find the dispersion relation for the surface electromagnetic waves propagating along the interface of a narrow-gap semiconductor occupying the half-space $x<0$ and a vacuum. The static electric field is assumed to be applied along the z-axis. When a constant electric current of density $\boldsymbol{J}_0 = \sigma(E_0^2)\boldsymbol{E}_0$ flows across the crystal (σ is the conductivity regarded as a function of the square of the field), a small perturbation (fluctuation) $\boldsymbol{E}$ in $\boldsymbol{E}_0$ leads to a change $\boldsymbol{J}$ in $\boldsymbol{J}_0$. The components of the vector $\boldsymbol{J}$ are $J_{x,y} = \sigma E_{x,y}$, $J_z = \sigma_{\mathrm{d}} E_z$, where

$$\sigma_{\mathrm{d}} = \sigma\left(1 + 2\frac{\mathrm{d}\ln\sigma}{\mathrm{d}\ln E_0^2}\right) \tag{10.1}$$

is the differential conductivity.

From Part 2 of this book we know that the surface waves correspond to the perturbations $\boldsymbol{E}$, $\boldsymbol{H} \sim \exp[\mathrm{i}(\boldsymbol{kr} - \omega t)]$ with real tangential components k_y and k_z of the wave vector and a complex normal component such that Im $k_x>0$ in the space region $x>0$ (vacuum) and Im $k_x<0$ for $x<0$. Linearizing the Maxwell equations with respect to small perturbations $\boldsymbol{E}$, $\boldsymbol{H}$ and $\boldsymbol{J}$ and solving the corresponding dispersion equation, we find for arbitrary values of k_y and k_z that the normal component of the wave vector in the region $x<0$ has two different values: k_{0x} and k_{ex}, where

Radiowaves and Polaritons in Anisotropic Media. Roland H. Tarkhanyan and Nikolaos K. Uzunoglu

ISBN: 3-527-40615-8

$$k_{ox}^2 = k_0^2 \varepsilon(\omega) - k_{||}^2 \tag{10.2}$$

$$k_{ex}^2 = k_0^2 \varepsilon_d(\omega) - k_y^2 - \frac{\varepsilon_d(\omega)}{\varepsilon(\omega)} k_z^2 \tag{10.3}$$

$$\varepsilon(\omega) = \varepsilon + \frac{i\sigma}{\varepsilon_0 \omega} \tag{10.4}$$

$$\varepsilon_d(\omega) = \varepsilon + \frac{i\sigma_d}{\varepsilon_0 \omega} \tag{10.5}$$

$$k_0^2 \equiv \frac{\omega^2}{c^2} \qquad k_{||}^2 \equiv k_y^2 + k_z^2 \tag{10.6}$$

and ε is the static permittivity of the crystal lattice. The wave with k_{ox} corresponds to the ordinary wave and the wave with k_{ex} to the extraordinary wave in the optics of uniaxial crystals; the preferred direction is represented by the direction of the static electric field. The normal component of the wave vector in the region $x > 0$ is given by

$$k_x^2 = k_0^2 - k_{||}^2 \tag{10.7}$$

Imposing standard boundary conditions of continuity of the fields at the interface plane $x = 0$, we obtain the following dispersion relation for the surface wave, which is a linear combination of the ordinary and extraordinary waves:

$$(k_x - k_{ox})\left\{k_{ox}k_{ex}[k_x - k_{ex}\varepsilon_d^{-1}(\omega)] + k_y^2[k_x - k_{ex} + k_{ox}(1 - \varepsilon_d^{-1}(\omega))]\right\} = 0 \tag{10.8}$$

Solutions of Eq. (10.8) give the complex angular frequency of surface waves as a nonexplicit function of the tangential wave numbers k_y and k_z. In the next two sections we shall examine these solutions in the absence as well as in the presence of retardation effects.

10.2
Stability of Surface Waves in the Absence of Retardation

Using Eqs. (10.2), (10.3) and (10.7), the dispersion relation (10.8) can be transformed to the equivalent form

$$k_z^2(k_x - k_{ox})[k_x - k_{ex} + k_{ox}(1 - \varepsilon^{-1}(\omega))] + k_0^2(k_x - k_{ex})[k_{ox} - \varepsilon(\omega)k_x] = 0 \tag{10.9}$$

In the absence of retardation ($c \to \infty$), Eq. (10.9) splits in two:

$$k_x - k_{ox} = 0 \tag{10.10}$$

and

$$k_x - k_{ex} + k_{ox}[1 - \varepsilon^{-1}(\omega)] = 0 \tag{10.11}$$

which characterize potential surface waves traveling in an arbitrary direction along the interface crystal–vacuum. From Eqs. (10.2) and (10.7} we find that

$$k_x = -k_{ox} = \mathrm{i}k_{||} \tag{10.12}$$

It follows that the ordinary surface wave described by the dispersion relation (10.10) does not exist. For the extraordinary potential surface wave, Eq. (10.11) yields

$$k_{ex} = \mathrm{i}k_{||}\varepsilon^{-1}(\omega) \tag{10.13}$$

which is fully consistent with the expression

$$k_{ex}^2 = -k_y^2 - \varepsilon_d(\omega)\varepsilon^{-1}(\omega)k_z^2 \tag{10.14}$$

and leads to the following dispersion relation:

$$\sqrt{\varepsilon(\omega)[\varepsilon(\omega)\sin^2\varphi + \varepsilon_d(\omega)\cos^2\varphi]} = -1 \tag{10.15}$$

where φ is the angle between $\boldsymbol{k}_{||}$ and $\boldsymbol{E}_0$. Solutions of Eq. (10.15) give two discrete complex values for ω, which depend on the direction of propagation, that is, on the angle φ. However, it is obvious that the growth of the amplitude of the surface oscillations described by Eq. (10.15) is impossible. In fact, since Im $k_{ex} < 0$ (this is the condition of existence of the surface waves), we find from Eq. (10.13) that Re$[\varepsilon^{-1}(\omega)] < 0$ or, using Eq. (10.4), we obtain

$$\varepsilon + \frac{\sigma \operatorname{Im} \omega}{\varepsilon_0 |\omega|^2} < 0 \tag{10.16}$$

This condition implies that Im $\omega < 0$; therefore, the wave amplitude decays exponentially in time. Consequently, the potential surface waves are stable and they cannot be amplified spontaneously.

10.3
Radiative Instability of Surface Electromagnetic Waves

For the waves traveling along the static electric field we have $k_y = 0$ and then Eq. (10.8) yields the following dispersion relation:

$$k_{ex} = \varepsilon_d(\omega)k_x \tag{10.17}$$

which characterizes the surface wave of TM type in the presence of retardation. As to the surface wave of TE type, it is easy to see that it does not exist in the presence of retardation, too.

Substituting Eqs. (10.3) and (10.7) for $k_y = 0$ into Eq. (10.17), we transform this equation to the form

$$\frac{c^2 k_{\parallel}^2}{\omega^2} = \frac{\varepsilon_{\rm d}(\omega) - 1}{\varepsilon_{\rm d}(\omega) - \varepsilon^{-1}(\omega)} \tag{10.18}$$

which is quartic in ω. In the frequency region where

$$\max\left\{\frac{\sigma}{\varepsilon_0 \omega}, \frac{|\sigma_{\rm d}|}{\varepsilon_0 \omega}\right\} << \operatorname{Re} \omega << \tau^{-1} \tag{10.19}$$

Eq. (10.18) has an approximate solution $\omega = \Omega + {\rm i}\gamma$, where $\Omega \gg \gamma$,

$$\Omega = ck_{\parallel}\sqrt{\varepsilon^{-1} + 1} \tag{10.20}$$

and

$$\gamma = \frac{\sigma - \varepsilon \sigma_{\rm d}}{2\varepsilon_0 \varepsilon(\varepsilon^2 - 1)} \tag{10.21}$$

The normal components of the wave vectors are, with accuracy up to terms $\sim\gamma/\omega$, given by

$$k_x = k_{\parallel}\varepsilon^{-1/2} + {\rm i}\gamma(\varepsilon + 1)^{1/2} c^{-1} \tag{10.22}$$

$$k_{\rm ex} = k_{\parallel}\varepsilon^{1/2} + {\rm i}[\gamma\varepsilon + \sigma_{\rm d}\varepsilon_0^{-1}(\varepsilon + 1)^{-1}]c^{-1} \tag{10.23}$$

Using the conditions Im $k_x > 0$ and Im $k_{\rm ex} < 0$, necessary for the existence of the surface wave, we obtain

$$\gamma > 0 \tag{10.24}$$

$$\sigma_{\rm d} < 0 \qquad |\sigma_{\rm d}| > \sigma(\varepsilon - 2)^{-1} \tag{10.25}$$

(it is assumed that $\varepsilon > 2$). Condition (10.24) describes the growth of the wave amplitude in time [$E \sim \exp(\gamma t)$], that is, it corresponds to an instability. Consequently, we find that, if the criterion (10.25) is satisfied in crystals with a negative differential conductivity, TM-type surface electromagnetic waves can be excited spontaneously. We have to remember that such a phenomenon is impossible for the bulk waves traveling in the same direction (see the end of Section 9.2).

It should also be noted that nonlinear effects obviously become important as the amplitude grows. If nonlinear terms are included in the relation $\boldsymbol{J} = \boldsymbol{J}(\boldsymbol{E})$, the growth of the amplitude becomes restricted and, as a result, a constant value of the amplitude is achieved under steady-state conditions. The growth increment γ then vanishes and the normal component of the wave vector in the region $x>0$ becomes real: $k_x = k_{\|}\varepsilon^{-1/2}$. It follows that the surface wave under steady-state conditions is emitted into vacuum and the energy lost by this radiation is supplied by a dc source.

The angle θ between $\boldsymbol{E}_0$ and the wave vector of the electromagnetic wave emanating from the crystal surface is given by

$$\tan\theta = k_x/k_{\|} = \varepsilon^{-1/2} \tag{10.26}$$

Setting, for example, $\varepsilon = 12.53$ (GaAs), we obtain $\theta = 15^o46$· Setting $k_{\|} = 2\pi l/L_z$, $l = 1, 2, \ldots$ and $L_z = 1cm$ as the length of the sample in the $\boldsymbol{E}_0$ direction, we obtain from Eq. (10.20) the following values for the frequencies of the excited waves: $\nu_l = \Omega/2\pi = 3.12x10^{10}l(s^{-1})$.

Thus, we find that the instability of surface waves manifests itself in the emission of microwaves due to the excitation of surface-emitted waves into the vacuum, and the crystal becomes a microwave oscillator [102].

It is worth noting that, in contrast to the waves considered in this section, the instability of the surface electromagnetic waves traveling in the direction perpendicular to $\boldsymbol{E}_0$ is impossible.

10.4 Nonradiative Instability of Interface Waves in Semiconductor Heterostructures

In this section we consider the instability of electromagnetic waves localized at the interface between narrow-gap and wide-gap semiconductor materials, and traveling along the static field $\boldsymbol{E}_0$ that is parallel to the z-axis. We continue our assumption that a narrow-gap material (for example, n-GaAs) occupies the half-space $x<0$; the region $x>0$ is now occupied by a wide-gap material (e.g. n-$Al_xGa_{1-x}As$). As we will see, both the dispersion relation and the instability criteria for the waves in a heterostructure are essentially different from those considered in the previous section. Moreover, in contrast to the radiative behavior of the instability, in heterostructures an instability of nonradiative interface waves can occur.

We shall only consider the extraordinary TM-type interface waves, for which the magnetic field vector $\boldsymbol{H}$ is parallel to the y-axis. Using the standard continuity conditions for the fields at the plane $x = 0$, we obtain the dispersion relation in the form

$$\frac{c^2k_{\|}^2}{\omega^2} = \frac{\varepsilon(\omega)\varepsilon_1(\omega)[\varepsilon_d(\omega) - \varepsilon_{1d}(\omega)]}{\varepsilon(\omega)\varepsilon_d(\omega) - \varepsilon_1(\omega)\varepsilon_{1d}(\omega)} \tag{10.27}$$

where $\varepsilon(\omega)$ and $\varepsilon_d(\omega)$ are given by Eqs. (10.4) and (10.5); $\varepsilon_1(\omega)$ and $\varepsilon_{1d}(\omega)$ are those for the wide-gap material, respectively.

In the frequency region (10.19), Eq. (10.27) has an approximate solution $\omega = \Omega + i\gamma$, where

$$\Omega \cong ck_{||}\sqrt{(\varepsilon + \varepsilon_1)(\varepsilon\varepsilon_1)^{-1}} \tag{10.28}$$

$$\gamma = [\varepsilon_1(\sigma\varepsilon_1\varepsilon^{-1} - \sigma_d) - \varepsilon(\sigma_1\varepsilon\varepsilon_1^{-1} - \sigma_{1d})]/2\varepsilon_0(\varepsilon^2 - \varepsilon_1^2) \tag{10.29}$$

and $\gamma << \Omega$. The normal components of the wave vector in the regions $x > 0$ and $x < 0$ are, with an accuracy up to terms $\sim\gamma/\Omega$, given by

$$k_{1x} = k_{||}\sqrt{\varepsilon_1\varepsilon^{-1}} + i\varepsilon[(\sigma_1\varepsilon_1^{-1} + \sigma_{1d}\varepsilon^{-1})(2\varepsilon_0)^{-1} + \gamma(1 + \varepsilon_1\varepsilon^{-1})]/c\sqrt{\varepsilon + \varepsilon_1} \tag{10.30}$$

$$k_x = k_{||}\sqrt{\varepsilon\varepsilon_1^{-1}} + i\varepsilon_1[(\sigma\varepsilon^{-1} + \sigma_d\varepsilon_1^{-1})(2\varepsilon_0)^{-1} + \gamma(1 + \varepsilon\varepsilon_1^{-1})]/c\sqrt{\varepsilon + \varepsilon_1} \tag{10.31}$$

Assuming that for the wide-gap material the value of σ_{1d} is positive, and using the conditions of existence of the interface waves $\mathrm{Im}\, k_{1x} > 0$ and $\mathrm{Im}\, k_x < 0$, from Eqs. (10.29)–(10.31) we obtain the following instability criteria:

$$\sigma_d < 0 \qquad |\sigma_d| > \sigma\varepsilon_1/\varepsilon \tag{10.32}$$

$$\sigma + |\sigma_d|(2 - \varepsilon\varepsilon_1^{-1}) < \sigma_1 - \varepsilon\sigma_{1d}\varepsilon_1^{-1} < |\sigma_d| + \sigma\varepsilon_1/\varepsilon \tag{10.33}$$

When these conditions are fulfilled, nonradiative interface waves can be excited spontaneously. Setting $\varepsilon_1 = 12.1$ and again $\varepsilon = 12.53$, $L_z = 1\,\text{cm}$, we obtain from Eq. (10.28) the following frequencies of the excited waves: $\nu_l = 1.2 \times 10^{10} l\,(\text{s}^{-1})$, $l = 1, 2, \ldots$.

Note that in the case when the conductivity of the wide-gap material can be neglected, the instability condition takes the form

$$\sigma_d < 0 \qquad |\sigma_d| > \sigma(\varepsilon\varepsilon_1^{-1} - 2)^{-1} \tag{10.34}$$

which for $\varepsilon_1 = 1$ coincides with Eq. (10.25). It is obvious that in this case the excited wave has a radiative behavior, that is, under the steady-state conditions it is emitted into the space region $x > 0$.

10.5 Constitutive Relations for Current Perturbations in the Presence of a Hot Two-Dimensional Electron Gas (2DEG)

In the previous section we did not take into account the presence of two-dimensional (2D) carriers localized at the interface between two different semiconductor materials. A two-dimensional electron gas (2DEG) is formed at a heterojunction GaAs–AlGaAs when the AlGaAs is n-doped while the GaAs is undoped. This is known as modulation doping [82]. The electrons will be thermally excited into the conduction band of AlGaAs. They will then migrate into the adjoining GaAs, since there they can achieve states of lower energy. The ionized donors in AlGaAs will form a net positive charge, which will attract the mobile electrons which are now in the GaAs, but do not have enough energy to recombine with their ionized donors. They will thus be trapped in the vicinity of the interface, since the band-edge discontinuity will prevent them from moving to the back, while the Coulomb attraction of the positive charge will keep them from moving far into the GaAs. This process continues until the system reaches equilibrium, and the mobile electrons have been trapped in a quantum well, forming the 2DEG. To investigate the influence of the 2DEG on the surface microwave excitation in a selectively doped heterostructure, we have to use the constitutive relations for the current perturbations in the presence of a nonlinear I–V characteristic for both the bulk and the 2D carriers. We assume that a negative differential conductivity of the 2DEG is caused by spatial transport of the carriers across the heterobarrier in a heating static electric field applied along the interface between the layers. Such a mechanism of NDR has been suggested in Refs. [95] and [108]. In Ref. [109] it has been shown experimentally that the threshold field $E^{\mathrm{s}}_{\mathrm{cr}}$ in this case is considerably smaller than that (E^{i}_{cr}) for intervalley transport of hot bulk carriers in the interior of the GaAs.

Now let us continue the consideration of the heterostructure described in the previous section assuming the presence of the 2DEG localized at the interface $x = 0$. In a static field $\boldsymbol{E}_0 || oz$, the 2DEG is characterized by the surface current density

$$\boldsymbol{J}_{0\mathrm{s}} = q n_{0\mathrm{s}} \boldsymbol{v}_{0\mathrm{s}} \tag{10.35}$$

where $n_{0\mathrm{s}} = n_{0\mathrm{s}}(E_0^2)$ is the surface density of the carriers,

$$\boldsymbol{v}_{0\mathrm{s}} = \mu_{\mathrm{s}} \boldsymbol{E}_0 \tag{10.36}$$

is the average drift velocity and $\mu_{\mathrm{s}} = \mu_{\mathrm{s}}(E_0^2)$ is the mobility. A small perturbation of the electric field $\boldsymbol{E}$ causes a change n_{s} in $n_{0\mathrm{s}}$, $\boldsymbol{v}_{\mathrm{s}}$ in $\boldsymbol{V}_{0\mathrm{s}}$ and $\boldsymbol{J}_{\mathrm{s}}$ in $\boldsymbol{J}_{0\mathrm{s}}$, where

$$\boldsymbol{v}_{\mathrm{s}} = \mu_{\mathrm{s}}^{\mathrm{d}} \boldsymbol{E} \tag{10.37}$$

$$\boldsymbol{J}_{\mathrm{s}} = q(n_{\mathrm{s}} \boldsymbol{v}_{0\mathrm{s}} + n_{0\mathrm{s}} \boldsymbol{v}_{\mathrm{s}}) \tag{10.38}$$

and μ_s^d is the two-dimensional tensor of the differential mobility. In the above-chosen system of coordinates it is a diagonal tensor:

$$(\mu_s^d)_{yy} = \mu_s, \ (\mu_s^d)_{zz} = \mu_s\left(1 + 2\frac{d\ln\mu_s}{d\ln E_0^2}\right) \equiv \mu_s^d \tag{10.39}$$

The perturbations of the surface charge and the current densities at the interface plane $x = 0$ are connected by the boundary condition

$$\sigma_1 E_{1x}^s - \sigma E_x^s + i(k_y J_{sy} + k_z J_{sz}) - i\omega q n_s = 0 \tag{10.40}$$

where E_x^s and E_{1x}^s are the normal components of the vector $\boldsymbol{E}$ at $x \to 0$ in the corresponding regions. Equation (10.40) can be derived from the equation of continuity

$$\operatorname{div}\boldsymbol{J} + \frac{\partial\rho}{\partial t} = 0 \tag{10.41}$$

by integration over a small volume containing the interface from $x = -\varepsilon$ to $x = +\varepsilon$ and then setting $\varepsilon \to 0$.

Substituting Eq. (10.38) into Eq. (10.40), for the surface charge density perturbation $\rho_s = q n_s$ we obtain

$$\rho_s = [q n_{0s} \boldsymbol{k}_{||} \boldsymbol{v}_s - i(\sigma_1 E_{1x}^s - \sigma E_x^s)]\varpi^{-1} \tag{10.42}$$

where

$$\varpi = \omega - k_z v_{0s} \tag{10.42a}$$

The first term on the right-hand side of Eq. (10.42) is caused by the 2D carriers, while the second term takes into account the charge flowing to the interface from the interior of the semiconductor materials by the normal components of the electric fields.

Using Eqs. (10.38) and (10.42), for the components of the 2D vector $\boldsymbol{J}_s$ we obtain

$$J_{sy} = \sigma_s E_y \tag{10.43}$$

$$J_{sz} = \{\omega\sigma_s^d E_z + v_{0s}[\sigma_s k_y E_y - i(\sigma_1 E_{1x}^s - \sigma E_x^s)]\}\varpi^{-1} \tag{10.44}$$

where

$$\sigma_s = q n_{0s}\mu_s, \qquad \sigma_s^d = q n_{0s}\mu_s^d \tag{10.45}$$

Analogously, in the narrow-gap material the current density

$$J_0 = \rho_0 v_0 = \rho_0 \mu E_0 \tag{10.46}$$

changes by the vector

$$J = \rho v_0 + \rho_0 v \tag{10.47}$$

where $\rho_0 = qn_0$ is the bulk charge density, $\mu = \mu(E_0^2)$ is the average mobility of the bulk carriers and ρ is the induced bulk charge density, which is related to the current density perturbation J by Eq. (10.41). The perturbation of the drift velocity is given by

$$v = \mu_d E \tag{10.48}$$

where the tensor μ_d has the form

$$\mu_d = \begin{pmatrix} \mu & 0 & 0 \\ 0 & \mu & 0 \\ 0 & 0 & \mu_d \end{pmatrix} \tag{10.49}$$

where

$$\mu_d = \mu\left(1 + \frac{2\,d\ln\mu}{d\ln E_0^2}\right) \tag{10.49a}$$

Using Eqs. (10.41) and (10.47)–(10.49), for the bulk charge and current densities we obtain

$$\rho = \frac{\sigma(k_x E_x + k_y E_y) + \sigma_d k_z E_z}{\omega - k_z v_0} \tag{10.50}$$

$$J_{x,y} = \sigma E_{x,y} \qquad J_z = \frac{\omega\sigma_d E_z + \sigma v_0(k_x E_x + k_y E_y)}{\omega - k_z v_0} \tag{10.51}$$

where

$$\sigma \equiv qn_0\mu \qquad \sigma_d \equiv qn_0\mu_d \tag{10.52}$$

Neglecting the nonlinearity in the I–V characteristic of the wide-gap material, for the corresponding current density perturbation in this region we obtain

$$J_{1x,y} = \sigma_1 E_{1x,y} \qquad J_{1z} = \frac{\sigma_1[\omega E_{1z} + v_{01}(k_{1x}E_{1x} + k_y E_{1y})]}{\omega - k_z v_{01}} \tag{10.53}$$

The constitutive relations given by Eqs. (10.43), (10.44), (10.51) and (10.53), together with the Maxwell equations, provide the investigation of both the bulk and interface waves in semiconductor heterostructures in the presence of the 2DEG and a static heating electric field parallel to the interface.

10.6
Excitation of Quasistatic Interface Waves in Heterostructures with a 2DEG

The influence of the 2DEG on the surface microwave instabilities is essentially considerable in the absence of the retardation effects. We recall that in the case considered in Section 10.2, the instability of the quasistatic waves was impossible. That is why in this section we will restrict ourselves by consideration of the quasi-static interface modes, which are described by the equations

$$\text{rot}\,\boldsymbol{E} = 0 \qquad \text{div}\,\boldsymbol{E} = \frac{\rho}{\varepsilon_0\varepsilon}$$

and by the equation of continuity given in Eq. (10.41).

We consider a heterostructure with the ideal 2DEG, in which there is no motion in the confining direction, and where we neglect interactions between the electrons. As we know, the quasistatic interface modes correspond to the solutions of Eqs. (10.54), for which the field $\boldsymbol{E}$ decays exponentially and tends to zero at $|x| \to \infty$. Using the boundary conditions at the plane $x = 0$:

$$E_y = E_{1y} \qquad E_z = E_{1z} \tag{10.55a}$$

$$\varepsilon_1 E_{1x} - \varepsilon E_x = \rho_s/\varepsilon_0 \tag{10.55b}$$

and Eqs. (10.42), (10.51) and (10.53), we obtain the dispersion relation for the quasistatic interface modes in the form

$$\varepsilon_s(\omega)k_x - \varepsilon_{1s}(\omega)k_{1x} + \varepsilon_{0s} = 0 \tag{10.56}$$

where

$$k_{1x} = ik_{\|} \qquad k_x^2 = -k_y^2 - k_z^2\,\tilde{\varepsilon}_d(\omega)\,\tilde{\varepsilon}^{-1}(\omega) \tag{10.57}$$

$$\tilde{\varepsilon}(\omega) = \varepsilon + i\sigma[\varepsilon_0(\omega - k_z v_0)]^{-1} \qquad \tilde{\varepsilon}_d(\omega) = \varepsilon + i\sigma_d[\varepsilon_0(\omega - k_z v_0)]^{-1} \tag{10.58}$$

$$\varepsilon_s(\omega) = \varepsilon + i\sigma\varepsilon_0^{-1}\varpi^{-1} \qquad \varepsilon_{1s}(\omega) = \varepsilon_1 + i\sigma_1\varepsilon_0^{-1}\varpi^{-1} \tag{10.59}$$

$$\varepsilon_{0s} = (\sigma_s k_y^2 + \sigma_s^d k_z^2)\varepsilon_0^{-1}\varpi^{-1} \tag{10.60}$$

and ϖ is given by Eq. (10.42a). We have to note that in the absence of the 2DEG, $\varepsilon_{0s} = 0$ and the imaginary part of ω, satisfying Eq. (10.56), is negative. It means that an excitation of surface modes is impossible even if $\sigma_d < 0$. For $\varepsilon_{0s} \neq 0$, but $k_z = 0$, that is for the waves traveling in the direction perpendicular to the field $\boldsymbol{E}_0$, an instability is impossible even when both σ_d and σ_s^d are negative. For an arbitrary direction of propagation, Eq. (10.56) is cubic in ω. When the nonlinearity in the I–V characteristic of the narrow-gap material can be neglected, that is, when $\sigma_d = \sigma$, Eq. (10.56) has an exact solution

$$\omega = k_{||}v_{0s} + i\gamma(\varphi) \tag{10.61}$$

where

$$\gamma(\varphi) = -[\sigma + \sigma_1 + k_{||}(\sigma_s \sin^2\varphi + \sigma_s^d \cos^2\varphi)][\varepsilon_0(\varepsilon + \varepsilon_1)]^{-1} \tag{10.62}$$

Here φ is the angle between $\boldsymbol{k}_{||}$ and $\boldsymbol{E}_0$. Using Eq. (10.62), we obtain the following instability criterion:

$$\sigma_s^d < 0 \qquad |\sigma_s^d| > \sigma_s \tan^2\varphi + (\sigma + \sigma_1)(k_{||}\cos^2\varphi)^{-1} \tag{10.63}$$

At a given value of $|\sigma_s^d|$, there can only be amplified the waves for which

$$k_{||} > k_{\min} = (\sigma + \sigma_1)|\sigma_s^d|^{-1} \tag{10.64}$$

and, consequently,

$$\mathrm{Re}\,\omega > k_{\min}v_{0s}$$

For a given value of $k_{||}$ in the region (10.64), there can only be amplified the waves whose direction of propagation is restricted by $\varphi < \varphi_0$, where

$$\tan^2\varphi_0 = \frac{k_{||}|\sigma_s^d| - \sigma - \sigma_1}{k_{||}\sigma_s + \sigma + \sigma_1} \tag{10.66}$$

The upper angle limit φ_0 increases monotonically with increasing $k_{||}$ from zero at $k_{||} = k_{\min}$ to the maximum value

$$\varphi_{\max} = \tan^{-1}\left(\sqrt{|\sigma_s^d|/\sigma_s}\right) \tag{10.67}$$

at $k_{||} \rightarrow \infty$. Setting, for example, $\sigma = 2.88 \times 10^{12}\mathrm{s}^{-1}$, $\sigma_1 = 1.44 \times 10^{11}\mathrm{s}^{-1}$, $\sigma_s = 5.76 \times 10^9 \mathrm{cm/s}$, and $|\sigma_s^d|/\sigma_s = 0.125$, we obtain $\mathrm{Re}\,\omega = 2 \times 10^{11}\mathrm{s}^{-1}$, $\varphi_{\max} = 20°$ and $\lambda_{\max} = 2\pi/k_{\min} = 15$ m km.

Thus, the presence of the 2DEG in selectively doped semiconductor heterostructures leads to the instability of the quasistatic interface waves whose phase velocity coincides with the drift velocity of 2D carriers. The waves propagate only in certain directions, and their upper wavelength limit depends on both the bulk carrier mobility and the 2D carrier differential mobility [105].

10.7 Influence of Hot 2D Carriers on Excitation of Guided Microwave Charge Density Oscillations

Wave instabilities in semiconductors are of great interest as a means for generating and amplifying various types of waves, including charge-density waves (CDWs), which have important applications in integrated-circuit microwave and waveguide technology [110]. In this section we investigate how hot 2D mobile electrons affect the instability of guided CDWs in a semiconductor film (n-GaAs) bounded on both sides by selectively doped layers of a wide-gap material (n-AlGaAs). The instability is caused by heating of bulk carriers in the layer of narrow-gap material. Crucial here is, as in the previous section, the presence of a descending region of the I–V characteristic, where the differential conductivity of the 2DEG is negative. This region is caused by spatial transport of carriers across the heterobarriers when a heating electric field is applied. In contrast to the interface modes, the phase velocity of the guided modes is determined by the drift velocity $\boldsymbol{v}_0$ of the bulk carriers in the central layer of the structure. We will show that when the drift velocity of the 2D carriers satisfies $v_{0s} \neq v_0$, a novel class of the microwave charge density oscillations is excited in a symmetric three-layer heterostructure. These modes propagate in the central layer, to which they are confined, and have the feature that for a special direction of propagation, their frequency and wavelength have definite fixed values which depend on the parameters of both bulk and 2D carriers. Moreover, the instability occurs only for a limited range of angles between the tangential wave vector $\boldsymbol{k}_{\|}$ and the field $\boldsymbol{E}_0$ and then only if the thickness of the waveguide core exceeds a definite minimum value.

So, we consider a narrow-gap semiconductor with a static conductivity σ and a dielectric constant ε, occupying the space region $|x| < d$. The selectively doped wide-gap material, characterized by the constant values σ_1 and ε_1, occupies the regions $x < -d$ and $x > d$. As the structure heats up by the static electric field, a nonlinear region characterized by a differential conductivity σ_d forms in the I–V characteristic of the central layer. We will again neglect the nonlinearity of the I–V characteristic of the wide-gap layers. The specific and differential conductivities of the 2DEG localized at the interface planes $x = \pm d$ are characterized by σ_s and $\sigma_\mathrm{s}^\mathrm{d}$.

A small perturbation $\boldsymbol{E}$ of the electric field will alter the charge density and currents for both the bulk and the 2D carriers. In the narrow-gap material the current density changes by the vector $\boldsymbol{J}$ given in Eq. (10.51), where J_z can be written in an equivalent form

$$J_z = \sigma_\mathrm{d} E_z + \rho v_0 \qquad (10.68)$$

The surface current density changes by the vector $\boldsymbol{J}_s$, with components given in Eqs. (10.43) and (10.44). The latter expression can also be written in the form

$$J_{sz} = \sigma_s^d E_z + \rho_s v_{0s} \tag{10.69}$$

where ρ_s is given by Eq. (10.42).

The guided modes correspond to the solutions of Eqs. (10.54) for which the field $\boldsymbol{E}$ decays exponentially for $|x| > d$ and tends to zero at infinity, while in the region $|x| < d$ the field is expressible in terms of trigonometric functions. Writing this field in the form

$$\boldsymbol{E} = -\text{grad}\psi \qquad \psi(x, y, z, t) = \psi(x)\exp[\mathrm{i}(k_y y + k_z z - \omega t)] \tag{10.70}$$

we then find from Eqs. (10.41) and (10.54) an expression for the space charge density perturbation $\rho(x)$ in the central layer (the exponential factor is omitted):

$$\rho(x) = \mathrm{i}(\omega - k_z v_0)^{-1}[\sigma\psi''(x) - (\sigma k_y^2 + \sigma_d k_z^2)\psi(x)] \tag{10.71}$$

and the scalar wave equation

$$\psi''(x) - k_{||}^2\psi(x) = -\frac{\rho(x)}{\varepsilon_0\varepsilon} \tag{10.72}$$

Substituting Eq. (10.71) into Eq. (10.72) and neglecting the charge-density perturbation for the wide-gap materials, we find the wave equation in the form

$$\psi''(x) = [k_y^2 + k_z^2\,\tilde{\varepsilon}_d(\omega)\,\tilde{\varepsilon}^{-1}(\omega)]\psi(x) \tag{10.73a}$$

for $|x| < d$ and

$$\psi''(x) = k_{||}^2\psi(x) \tag{10.73b}$$

for $|x| > d$, where $\tilde{\varepsilon}(\omega)$ and $\tilde{\varepsilon}_d(\omega)$ are given by Eqs. (10.58). The solution of Eqs. (10.73a) and (10.73b) is given by

$$\psi(x) = \begin{cases} A\mathrm{e}^{-k_{||}x}, & x > d \\ B\mathrm{e}^{\mathrm{i}k_x x} + C\mathrm{e}^{-\mathrm{i}k_x x}, & |x| < d \\ D\mathrm{e}^{k_{||}x}, & x < -d \end{cases} \tag{10.74}$$

where k_x is determined in Eq. (10.57). To find the constants A, B, C and D, we use the boundary conditions

$$\psi(d+0) = \psi(d-0) \qquad \psi(-d+0) = \psi(-d-0) \tag{10.75}$$

$$\varepsilon\psi'(d-0) - \varepsilon_1\psi'(d+0) = \varepsilon_0^{-1}\rho_s(d) \qquad \varepsilon_1\psi'(-d-0) - \varepsilon\psi'(-d+0) = \varepsilon_0^{-1}\rho_s(-d) \tag{10.76}$$

which express the continuity of the electric field tangential components E_y, E_z and the discontinuity of the normal component D_x of the electric displacement vector at the layer boundaries $x = d$ and $x = -d$, respectively. Expressions for the surface charge density perturbations $\rho_s(\pm d)$ can be found by integrating the continuity equation (10.41) along an infinitesimally thin layer at the corresponding interface and recalling Eqs. (10.43) and (10.69). After some straightforward algebra, we then obtain

$$\rho_s(\pm d) = -\mathrm{i}\varpi^{-1}\{(\sigma_s k_y^2 + \sigma_s^d k_z^2)\psi(\pm d) \pm [\sigma\psi'(\pm(d-0)) - \sigma_1\psi'(\pm(d+0))]\} \tag{10.77}$$

Inserting Eqs. (10.74) and (10.77) into Eqs. (10.75) and (10.76), we obtain a system of four linear homogeneous equations for the coefficients A, B, C and D. The condition for these equations to have a simultaneous solution leads to two dispersion equations relating the wave frequency ω and the tangential wave vector $\boldsymbol{K}_{||} = \{k_y, k_z\}$:

$$\frac{\varepsilon_s(\omega)k_x}{\mathrm{i}\varepsilon_{0s} + k_{||}\varepsilon_{1s}(\omega)} = \begin{cases} \cot(k_x d) \\ -\tan(k_x d) \end{cases} \tag{10.78a, b}$$

where $\varepsilon_s(\omega)$, $\varepsilon_{1s}(\omega)$ and ε_{0s} are given by Eqs. (10.59) and (10.60).

Equations (10.78a) and (10.78b) correspond to even $[B = C,\ \psi(x) \sim \cos(k_x x)]$ and odd modes $[B = -C,\ \psi(x) \sim \sin(k_x x)]$, respectively. According to Eq. (10.57), the guided modes can only exist (i.e. k_x is real) if $\sigma_d \neq \sigma$, that is only if the bulk carriers are heated in the region $|x| < d$. In addition, the conditions

$$\mathrm{Im}[\tilde{\varepsilon}_d(\omega)\,\tilde{\varepsilon}^{-1}(\omega)] = 0 \tag{10.79}$$

$$\tilde{\varepsilon}_d(\omega)\,\tilde{\varepsilon}^{-1}(\omega) + \tan^2\varphi < 0 \tag{10.80}$$

must be satisfied, where φ is the angle between the static field $\boldsymbol{E}_0$ and the vector $\boldsymbol{k}_{||}$. From Eq. (10.79), we obtain

$$\mathrm{Re}\,\omega = k_{||}v_0\cos\varphi \tag{10.81}$$

which implies that the phase velocity of the guided modes is determined by the average drift velocity v_0 of the bulk carriers in the narrow-gap material and depends on the direction of propagation relative to the field $\boldsymbol{E}_0$. Thus, in a symmetric heterostructure, for specified values of $k_{||}$ and φ, the even and odd modes have the same frequency $\mathrm{Re}\omega$ but different growth (or damping) rates $\gamma = \mathrm{Im}\,\omega$, which are given by the solutions of Eqs. (10.78a) and (10.78b). Inserting Eq. (10.81) into Eq. (10.80), we find that the guided modes can only be unstable if the conditions

$$\sigma_{\rm d} < 0 \qquad |\sigma_{\rm d}| > \sigma \tan^2\varphi \tag{10.82}$$

$$0 < \gamma < \gamma_{\rm cr} \tag{10.83}$$

are satisfied, where

$$\gamma_{\rm cr} = (|\sigma_{\rm d}| - \sigma \tan^2\varphi)[\varepsilon_0 \varepsilon (1 + \tan^2\varphi)]^{-1} \tag{10.84}$$

Since the threshold for negative differential conductivity for 2D carriers is small compared to that for bulk carriers, when Eq. (10.82) holds we have the further condition $\sigma_{\rm s}^{\rm d} < 0$.

Turning now to the analysis of Eqs. (10.78a) and (10.78b), we substitute $k_x = a k_{||}$ and use Eqs. (10.57)–(10.60) and (10.81) to obtain

$$\frac{a\beta\varepsilon}{\varepsilon_1} = \begin{cases} \cot(ak_{||}d) \\ -\tan(ak_{||}d) \end{cases} \tag{10.85a, b}$$

for even and odd modes, respectively, where

$$a^2 = (|\sigma_{\rm d}| + \sigma)(\sigma + \varepsilon_0\varepsilon\gamma)^{-1}\cos^2\varphi - 1 \tag{10.86}$$

$$\beta = \frac{k_z(v_0 - v_{0\rm s}) + {\rm i}(\gamma + \sigma\varepsilon_0^{-1}\varepsilon^{-1})}{k_z(v_0 - v_{0\rm s}) + {\rm i}\{\gamma + [\sigma_1 + k_{||}(\sigma_{\rm s}\sin^2\varphi + \sigma_{\rm s}^{\rm d}\cos^2\varphi)]\varepsilon_0^{-1}\varepsilon^{-1}\}} \tag{10.87}$$

The instability condition (10.83) now becomes

$$0 < a < \sqrt{\varepsilon_0 \varepsilon \gamma_{\rm cr}} \tag{10.88}$$

Since the right-hand sides of Eqs. (10.85a) and (10.85b) are real, we must set Imβ $= 0$. This condition holds in only two cases: (1) $v_0 = v_{0\rm s}$; (2) $v_0 \neq v_{0\rm s}$, but

$$\sigma = \varepsilon\varepsilon_1^{-1}[\sigma_1 + k_{||}(\sigma_{\rm s}\sin^2\varphi + \sigma_{\rm s}^{\rm d}\cos^2\varphi)] \tag{10.89}$$

Since $v_{0\rm s} \neq v_0$ for the heterostructures encountered in practice (usually $v_{0\rm s} > v_0$), we will henceforth consider only the situation when Eq. (10.89) holds. Then, in Eqs. (10.85a) and (10.85b) we must set $\beta = 1$, and

$$k_{||} = (\sigma\varepsilon_1\varepsilon^{-1} - \sigma_1)(\sigma_{\rm s}\sin^2\varphi + \sigma_{\rm s}^{\rm d}\cos^2\varphi)^{-1} \equiv \tilde{k} \tag{10.90}$$

In our heterostructure we have $\sigma\varepsilon_1 > \sigma_1\varepsilon$, $\sigma_{\rm s}^{\rm d} < 0$, and, since $k_{||}$ is the modulus of the vector $\boldsymbol{K}_{||}$, we see from Eq. (10.90) that an unstable wave can only exist if

$$|\sigma_{\rm s}^{\rm d}| < \sigma_{\rm s}\tan^2\varphi \tag{10.91}$$

Combining this condition with Eq. (10.82), we conclude that such a wave can only be excited for a restricted range of angles φ:

$$|\sigma_s^d|/\sigma_s < \tan^2\varphi < |\sigma_d|/\sigma \tag{10.92}$$

As φ increases in this interval, $\tilde{k}$ decreases monotonically but always remains greater than the minimum value given by

$$k_0 = (\sigma + |\sigma_d|)(\sigma\varepsilon_1 - \sigma_1\varepsilon)\varepsilon^{-1}(\sigma_s|\sigma_d| - \sigma|\sigma_s^d|)^{-1} \tag{10.93}$$

This means that the wavelength of the excited mode is bounded from above and the frequency from below: $\lambda < \lambda_{\max}$, $\mathrm{Re}\,\omega > \omega_{\min}$, where

$$\lambda_{\max} = 2\pi/k_0 \qquad \omega_{\min} = k_0 v_0(1 + |\sigma_d|\sigma^{-1})^{-1/2} \tag{10.94}$$

Thus, when $v_{0s} \neq v_0$, unstable guided space charge modes can be excited with definite values of the wave number (10.90) and frequency (10.81), which depend on the parameters of both bulk and 2D carriers [111]. These values are identical for the even and odd modes and are independent of the thickness of the active central layer. The guided modes are only excited when both bulk and 2D carriers exhibit negative differential conductivity, and in this case the phase velocity depends on the direction of the wave propagation and is bounded by

$$v_0(1 + \sigma^{-1}|\sigma_d|)^{-1/2} < v_{ph} < v_0(1 + \sigma_s^{-1}|\sigma_s^d|)^{-1/2} \tag{10.95}$$

Unlike the frequency and wavelength, the growth rate depends strongly on the central layer thickness $L_x = 2d$ and is different for even and odd modes. The dependence of γ on L_x is given by the transcendental equations

$$\frac{a\varepsilon_1}{\varepsilon} = \begin{cases} \cot(a\tilde{k}d) \\ -\tan(a\tilde{k}d) \end{cases} \tag{10.96}$$

which have a set of distinct discrete solutions $a_n = a_n(\tilde{k}d)$, each corresponding to a particular branch of the multivalued function

$$\gamma_n(\tilde{k}d) = \frac{1}{\varepsilon_0\varepsilon}\left[-\sigma + \frac{(\sigma + |\sigma_d|)\cos^2\varphi}{1 + a_n^2}\right] \tag{10.97}$$

where the subscript $n = 0, 1, 2, \ldots$ is the number of the branch. These branches do not cross anywhere. Each branch corresponds to a particular "threshold" value of the thickness L_n of the waveguide core at which instability begins, and the higher the branch number n, the larger the corresponding threshold L_n. The instability condition can thus be expressed in the form

$$L_x > L_n \qquad L_n = \frac{2}{\tilde{a}\,\tilde{k}} \begin{cases} (\pi n + \delta) \\ [\pi(n + 1/2) + \delta] \end{cases} \tag{10.98}$$

$$\tilde{a} = (|\sigma_d| - \sigma \tan^2\varphi)^{1/2} \cos\varphi \qquad \cot\delta = \varepsilon\,\tilde{a}\,\varepsilon_1^{-1} \tag{10.99}$$

The increase in wave amplitude with time can be characterized in a natural way by that branch γ_n which has the smallest positive value for the given thickness L_x. It means that the growth rate has a "dimensionally piecewise" character: $\gamma = \gamma_0$ if $L_0 < L_x \le L_1$, $\gamma = \gamma_1$ if $L_1 < L_x \le L_2$ and, in general, $\gamma = \gamma_n$ if $L_n < L_x \le L_{n+1}$. All the branches γ_n are monotonically increasing functions of L_x, and the maximum value of γ_n decreases with increasing branch number (see Fig. 10.1). A guided mode with a specific wave number $\tilde{k}$ can thus be unstable only if the waveguide core thickness L_x exceeds a definite minimum value L_0. It is noteworthy that for excitation of an odd mode, the core must be more than twice as thick as for excitation of an even mode:

$$L_0^{\mathrm{odd}} / L_0^{\mathrm{even}} = 1 + \pi/2\delta > 2 \tag{10.100}$$

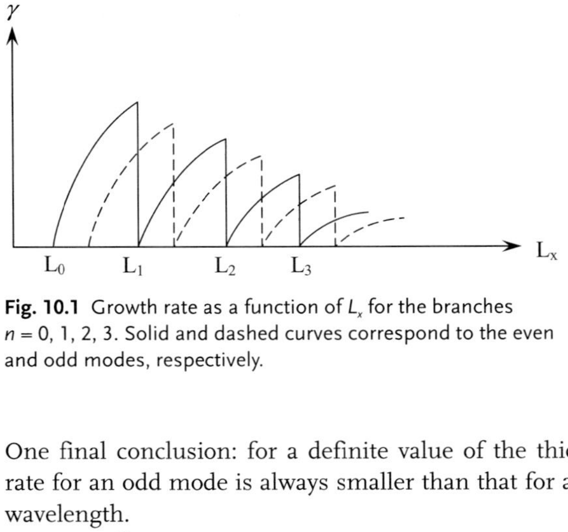

Fig. 10.1 Growth rate as a function of L_x for the branches $n = 0, 1, 2, 3$. Solid and dashed curves correspond to the even and odd modes, respectively.

One final conclusion: for a definite value of the thickness $L_x > L_0^{\mathrm{odd}}$, the growth rate for an odd mode is always smaller than that for an even mode with the same wavelength.

For small positive γ, i.e. near $L_x = L_n$, we can set $a = \tilde{a}$ in the left-hand side of Eqs. (10.96) and obtain

$$a_n \cong \tilde{a} L_n L_x^{-1} \tag{10.101}$$

where L_n is given by Eq. (10.98). Substituting Eq. (10.101) into Eq. (10.97), we obtain an explicit expression for γ_n as a function of L_x and φ :

$$\gamma_n = (\varepsilon_0\varepsilon)^{-1}(|\sigma_d|\cos^2\varphi - \sigma\sin^2\varphi)(L_x^2 - L_n^2)(L_x^2 + \tilde{a}^2 L_n^2)^{-1} \tag{10.102}$$

which again leads to instability conditions which, as might be expected, coincide precisely with Eqs. (10.98) and (10.82). On the other hand, far from $L_x = L_n$ (for example, if the core is sufficiently thick, so that $L_x\tilde{k} >> 1$) we find from Eqs. (10.96) and (10.97) a somewhat different expression for $\gamma_{.n}(L_x)$:

$$a_n \approx \frac{2\pi N}{\tilde{k}L_x} << 1 \tag{10.103}$$

$$\gamma_n = \frac{1}{\varepsilon_0\varepsilon}(|\sigma_d|\cos^2\varphi - \sigma\sin^2\varphi)\left[1 - \left(\frac{l_n}{L_x}\right)^2\right] \tag{10.104}$$

where

$$l_n = \frac{2\pi N}{\tilde{k}\,\tilde{a}}\sqrt{1+\tilde{a}^2} \tag{10.105}$$

Here $N = n + 1/2$ for even modes, and $N = n + 1$ for odd modes.

Since the normal component of the wave vector is given by $k_x = a_n\tilde{k}$ in the region $|x| < d$, in both cases considered above the phases of the odd and even modes differ by $\pi x/L_x$.

It should be noted that the waves described above can be excited not only in GaAs–AlGaAs structures but also in other three-layer heterostructures. The material must satisfy the following basic requirements: (1) mobile 2D carriers must be present on the interfaces between the layers; (2) both the bulk carriers in the central layer and the 2D carriers must exhibit negative differential conductivity; (3) the threshold field for the 2D carriers must be small compared to the field at which the bulk carrier current starts to decrease.

Part 4
Radiation of a Dipole Source in the Presence of a Grounded Gyromagnetic Dielectric Medium

Nikolaos K. Uzunoglu

Radiowaves and Polaritons in Anisotropic Media. Roland H. Tarkhanyan and Nikolaos K. Uzunoglu

ISBN: 3-527-40615-8

Introduction

In recent years, there is a growing interest in using anisotropic materials in optical signal processing and, in particular, to modify the radar cross section of scatterers. In a variety of applications, such as geophysical prospecting [112–114], remote sensing [115] and microstrip circuits and antennas [116–119], it is necessary to know the response of an anisotropic medium to an elementary source excitation, that is, the Green's function to compute the electromagnetic fields. Numerous authors have derived dyadic Green's functions for layered media, both isotropic and anisotropic [120–131]. The propagation properties of electromagnetic fields in gyromagnetic media have been used widely in the development of nonreciprocal microwave devices. The widespread use of printed circuits in microwave technology has led to interest in gyromagnetic materials in the past three decades [132–142]. The electromagnetic fields produced by an impressed field source in an infinitely extended gyromagnetic medium have been examined by Bunkin [132]. The problem of radiation by an elementary dipole in an infinite magneto-ionic medium and in the presence of a grounded gyroelectric plasma slab has been investigated by Arbel and Felsen [133] and by Wy [134]. In Ref. [134] the dipole axis is assumed to coincide with the axis of anisotropy.

In Chapter 11, the fundamental problem of the radiation of a dipole source in the presence of a grounded gyromagnetic slab will be examined [135]. A Fourier analysis is employed to compute the field, which is represented by using a dyadic formulation. The Green's function for an arbitrarily oriented dipole source is derived, assuming that the static magnetic field applied to the slab is either perpendicular or parallel to the slab surface. It is shown that the far-field amplitude and phase can be controlled by varying the external magnetic field. Numerical results are presented for several cases.

Radiowaves and Polaritons in Anisotropic Media. Roland H. Tarkhanyan and Nikolaos K. Uzunoglu

ISBN: 3-527-40615-8

11
Radiation of a Dipole in the Presence of a Grounded Gyromagnetic Slab

11.1
Formulation of the Problem

Consider a lossless homogeneous gyromagnetic material which is biased by an externally applied static magnetic field $\boldsymbol{H}_0$. The geometry of the problem is shown in Fig. 11.1. The primary excitation is an elementary dipole located in the region $z > 0$ above the gyromagnetic substrate (region 1) which is placed on an infinite perfect conductor plane at $z = -d$. The half-space $z > 0$ is assumed to be free (region 2). In the following analysis an $\exp(i\omega t)$ time dependence of the fields is assumed.

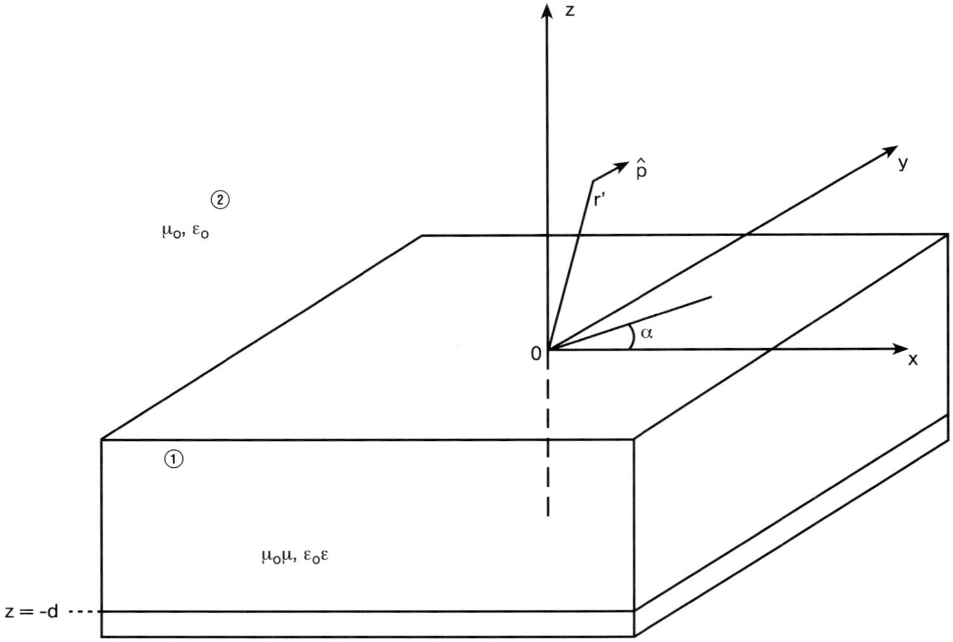

Fig. 11.1 The geometry of a radiating dipole in the presence of a gyromagnetic substrate.

Radiowaves and Polaritons in Anisotropic Media. Roland H. Tarkhanyan and Nikolaos K. Uzunoglu

ISBN: 3-527-40615-8

The ferrite slab (region 1) is characterized by a relative dielectric constant ε and a magnetic permeability tensor $\boldsymbol{\mu} = \boldsymbol{\mu}_{\perp}$ in the case of a perpendicular biasing magnetic field $\boldsymbol{H}_0 || z$ and by the tensor $\boldsymbol{\mu} = \boldsymbol{\mu}_{||}$ in the case of a parallel-directed magnetic field $\boldsymbol{H}_0 || y$, where

$$\boldsymbol{\mu}_{\perp} = \begin{pmatrix} \mu_1 & -\mathrm{i}\mu_2 & 0 \\ \mathrm{i}\mu_2 & \mu_1 & 0 \\ 0 & 0 & 1 \end{pmatrix} \tag{11.1}$$

and

$$\boldsymbol{\mu}_{||} = \begin{pmatrix} \mu_1 & 0 & \mathrm{i}\mu_2 \\ 0 & 1 & 0 \\ -\mathrm{i}\mu_2 & 0 & \mu_1 \end{pmatrix} \tag{11.2}$$

Here μ_1 is given in Eq. (8.1a) and μ_2 can be obtained from Eq. (8.1b) by changing the sign of ω. Note that the particular choice of the anisotropy axis in Eq. (11.2) does not reduce the generality of the treatment, since the direction of propagation is taken to be arbitrary in the xy-plane.

Maxwell's equations for the region 1 are

$$\text{rot}\,\boldsymbol{E}_1 = -\mathrm{i}\mu_0\omega\boldsymbol{\mu}\boldsymbol{H}_1 \qquad \text{rot}\,\boldsymbol{H}_1 = \mathrm{i}\omega\varepsilon_0\varepsilon\boldsymbol{E}_1 \tag{11.3}$$

and for the isotropic region 2

$$\text{rot}\,\boldsymbol{E}_2 = -\mathrm{i}\mu_0\omega\boldsymbol{H}_2 \qquad \text{rot}\,\boldsymbol{H} = \mathrm{i}\omega\varepsilon_0\boldsymbol{E}_2 + \boldsymbol{J} \tag{11.4}$$

where

$$\boldsymbol{J} = \boldsymbol{p}\delta(\boldsymbol{r} - \boldsymbol{r}') \qquad |\boldsymbol{p}| = 1 \tag{11.5}$$

is the elementary dipole source located at $\boldsymbol{r} = \boldsymbol{r}'$ with an arbitrary orientation of the strength $\boldsymbol{p}$.

The planar geometry of the boundary value problem suggests that a plane-wave Fourier representation for the electromagnetic fields would be appropriate. Consider first the slab region 1 and define the Fourier transformation

$$\boldsymbol{E}_1(\boldsymbol{r}) = (2\pi)^{-3/2} \int_{-\infty}^{+\infty} \mathrm{d}\boldsymbol{k}\boldsymbol{E}_1(\boldsymbol{k})\exp(-\mathrm{i}\boldsymbol{k}\boldsymbol{r}) \tag{11.6}$$

and the analogous expression for $\boldsymbol{H}_1(\boldsymbol{r})$. Eliminating the $\boldsymbol{E}_1$ field from Eqs. (11.3), a wave equation for the magnetic field is obtained:

$$\text{rot rot}\,\boldsymbol{H}_1 - k_1^2\boldsymbol{\mu}\boldsymbol{H}_1 = 0 \tag{11.7}$$

where

$$k_1 = k_0\sqrt{\varepsilon} \qquad k_0 = \omega/c \tag{11.8}$$

Using the Fourier transformation and requiring a nontrivial solution for the magnetic field, we obtain

$$\det(\delta_{ij}k^2 - k_i k_j - k_1^2 \mu_{ij}) = 0$$

Taking as unknown the $k_z = k\hat{z}$ term, where $\hat{z}$ is the unit vector along the z-axis, the solution of this biquadratic equation can be written in the form

$$k_z^2 = \frac{1}{2}\left(-S_1 \pm \sqrt{S_1^2 + S_2^2}\right) = \begin{cases} \tau_1^2 \text{ for } (+) \text{ sign} \\ \tau_2^2 \text{ for } (-) \text{ sign} \end{cases} \tag{11.10}$$

In the case of the perpendicular magnetization $\boldsymbol{H}_0 || \hat{z}$, the $S_{1,2}$ parameters are found to be

$$S_1 = 2\mu_1 k_1^2 - (\mu_1 + 1)k_t^2 \qquad > k_t^2 \equiv k_x^2 + k_y^2 \tag{11.11}$$

$$S_2 = 4[(\mu_1^2 - \mu_2^2)k_1^2 - \mu_1 k_t^2](k_t^2 - k_1^2) \tag{11.12}$$

while in the case of the parallel magnetization $\boldsymbol{H}_0 || \boldsymbol{y}$

$$S_1 = \mu_1^{-1}[(\mu_1^2 - \mu_2^2)k_1^2 + \mu_1(k_1^2 - k_t^2 - k_x^2) - k_y^2] \tag{11.13}$$

$$S_2 = 4\mu_1^{-2}\{[\mu_1 k_t^2 - (\mu_1^2 - \mu_2^2)][\mu_1(k_1^2 - k_x^2) - k_y^2] + (\mu_2 k_1 k_y)^2\} \tag{11.14}$$

The result of Eq. (11.10) states that, in order that the $\boldsymbol{E}_1$, $\boldsymbol{H}_1$ fields satisfy Maxwell's equations, the integration over the k_z variable in Eq. (11.6) should be reduced to a summation that includes only the terms $k_z = \tau_1$ and τ_2. According to this, Eq. (11.6) can be written in the form

$$\boldsymbol{E}_1(\boldsymbol{r}) = (2\pi)^{-1} \iint \mathrm{d}\boldsymbol{k}_t \boldsymbol{E}_1(\boldsymbol{k}_t, z)\exp(-\mathrm{i}\boldsymbol{k}_t \boldsymbol{r}) \tag{11.15}$$

[and an analogous expression for $\boldsymbol{H}_1(\boldsymbol{r})$], where $\boldsymbol{k}_t = k_x \hat{x} + k_y \hat{y}$,

$$\boldsymbol{E}_1(\boldsymbol{k}_t, z) = \sum_{j=1}^{2} [\boldsymbol{A}_j \cosh(\tau_j z) + \boldsymbol{B}_j \sinh(\tau_j z)] \tag{11.16a}$$

$$\boldsymbol{H}_1(\boldsymbol{k}_t, z) = \sum_{j=1}^{2} [\boldsymbol{C}_j \cosh(\tau_j z) + \boldsymbol{D}_j \sinh(\tau_j z)], \tag{11.16b}$$

Here $\boldsymbol{A}_j$, $\boldsymbol{B}_j$, $\boldsymbol{C}_j$, $\boldsymbol{D}_j$ are unknown coefficients to be determined. In the following consideration, it is convenient to examine separately the cases when the static magnetic field is perpendicular and parallel to the slab surface.

11.2
Dyadic Green's Function for Perpendicular Magnetization

Before proceeding further, let us decompose the electromagnetic fields into transverse and longitudinal components: $\boldsymbol{H}_1(\boldsymbol{r}) = \boldsymbol{H}_{1t} + \hat{z}\,H_{1z}$, $\boldsymbol{E}_1(\boldsymbol{r}) = \boldsymbol{E}_{1t} + \hat{z}\,E_{1z}$. The same decomposition holds for the transformed fields $\boldsymbol{E}_1(\boldsymbol{k}_t, z)$ and $\boldsymbol{H}_1(\boldsymbol{k}_t, z)$. Then, substituting Eqs. (11.15) and (11.16b) into Eq. (11.7) and solving for the transverse magnetic field, one obtains

$$\boldsymbol{H}_{1t}(\boldsymbol{k}_t, z) - \sum_{j=1}^{2} \hat{M}^{-1}(\tau_j)\boldsymbol{w}\tau_j h_{jz}(z) \tag{11.17}$$

where

$$h_{jz} = \frac{\mathrm{d}h^*_{jz}}{\mathrm{d}(\tau_j z)} \qquad h^*_{jz} = C_{jz}\cos(\tau_j z) + D_{jz}\sin(\tau_j z) \tag{11.18}$$

$$\boldsymbol{w} \equiv -\mathrm{i}\boldsymbol{k}_t \tag{11.19}$$

and $\hat{M}(\tau)$ is a 2 × 2 dyadic defined as

$$\hat{M}(\tau) = (k_y^2 - \tau^2 - \mu_1 k_1^2)\,\hat{x}\,\hat{x} + (\mathrm{i}\mu_2 k_1^2 - k_x k_y)\,\hat{x}\,\hat{y} - (\mathrm{i}\mu_2 k_1^2 + k_x k_y)\,\hat{y}\,\hat{x} + (k_x^2 - \tau^2 - \mu_1 k_1^2)\,\hat{y}\,\hat{y} \tag{11.20}$$

The longitudinal magnetic field is obtained directly from Eq. (11.16b):

$$H_{1z}(\boldsymbol{k}_t, z) = h^*_{1z}(z) + h^*_{2z}(z) \tag{11.21}$$

The two independent solutions h^*_{1z} and h^*_{2z} correspond to the extraordinary and ordinary waves for the anisotropic medium.

In order to express the $\boldsymbol{E}_1$ field in terms of $\boldsymbol{H}_1$, we substitute Eqs. (11.17) and (11.21) into the Maxwell–Ampere equation (11.3). After some manipulations, the result can be written as

$$\boldsymbol{E}_{1t}(\boldsymbol{k}_t, z) = -\mathrm{i}(\omega\varepsilon_0\varepsilon)^{-1}\sum_{j=1}^{2}[\boldsymbol{g} - \hat{N}(\tau_j)\,\hat{M}^{-1}(\tau_j)\boldsymbol{w}\tau_j]h^*_{jz} \tag{11.22}$$

$$E_{1z}(\boldsymbol{k}_t z) = \mathrm{i}(\omega\varepsilon_0\varepsilon)^{-1}\boldsymbol{g}\boldsymbol{H}_{1t}(\boldsymbol{k}_t, z) \tag{11.23}$$

where

$$\boldsymbol{g} = \boldsymbol{w}\boldsymbol{x}\hat{\boldsymbol{z}} \tag{11.24}$$

and

$$\hat{N}(\tau) = -\tau(\hat{x}\,\hat{y} - \hat{y}\,\hat{x}). \tag{11.25}$$

The fields in the isotropic region 2 can also be expressed in a similar manner: the primary fields **E**, **H** due to the elementary current excitation (11.5) are given from the well-known free-space dyadic Green's function [143]

$$\boldsymbol{E}(\boldsymbol{r}) = -\mathrm{i}(4\pi\varepsilon_0\omega)^{-1}(k_0^2\,\hat{I} + \vec{\nabla}\,\vec{\nabla})\boldsymbol{p}\frac{\exp(-\mathrm{i}k_0|\,\vec{r} - \vec{r}\,'|)}{|\,\vec{r} - \vec{r}\,'|} \tag{11.26a}$$

$$\boldsymbol{H}(\boldsymbol{r}) = \mathrm{i}(\omega\mu_0)^{-1}\mathrm{rot} \tag{11.26b}$$

where

$$(\hat{I}\,\boldsymbol{p})_\mathrm{i} = \delta_{ij}p_j = p_i \qquad (\vec{\nabla}\,\vec{\nabla}\,\boldsymbol{p}_i = \nabla_i\nabla_j p_j. \tag{11.26c}$$

Thus, the electric field may be viewed as a distribution; namely,

$$\boldsymbol{E}(\boldsymbol{r}) = -\mathrm{i}\omega\mu_0 G(\boldsymbol{r},\boldsymbol{r}')\boldsymbol{p} \tag{11.27}$$

where the dyadic Green's function is given by

$$G(\boldsymbol{r},\boldsymbol{r}') = (\hat{I} + k_0^{-2}\vec{\nabla}_r\vec{\nabla}_r)\frac{\exp(-ik_0R)}{4\pi R},\ R = |\boldsymbol{r} - \boldsymbol{r}'| \tag{11.28}$$

and $\hat{I}$ is the unit dyadic tensor.

Substituting the Fourier transform of the scalar function

$$\frac{\exp(-ik_0R)}{4\pi R} = \frac{1}{(2\pi)^3}\lim_{\varepsilon\to 0}\int_{-\infty}^{\infty} \mathrm{d}\boldsymbol{k}\frac{\exp[-\mathrm{i}\,\vec{k}\,(\vec{r} - \vec{r}\,')]}{k^2 - k_0^2 + \mathrm{i}\varepsilon} \tag{11.29}$$

into Eq. (11.27) and integrating for the k_z variable, it is possible to express the fields in the form of two-dimensional Fourier integrals:

$$\boldsymbol{E}(\boldsymbol{r}) = -\mathrm{i}(8\pi^2\varepsilon_0\omega)^{-1}\iint \mathrm{d}\boldsymbol{k}_t A(\boldsymbol{k}_t)(k_0^2\,\hat{I} - \boldsymbol{k}_v\boldsymbol{k}_v)\boldsymbol{p} \tag{11.30}$$

$$\boldsymbol{H}(\boldsymbol{r}) = (8\pi^2)^{-1}\iint \mathrm{d}\boldsymbol{k}_t A(\boldsymbol{k}_t)[\boldsymbol{k}_t x\boldsymbol{p}] \tag{11.31}$$

where

$$A(\boldsymbol{k}_t) = v_0^{-1}\exp[-v_0(z - z') - \mathrm{i}\boldsymbol{k}_t(\boldsymbol{r} - \boldsymbol{r}')] \quad z \neq z' \tag{11.32a}$$

$$v_0 = \sqrt{k_t^2 - k_0^2}\ \boldsymbol{k}_v = \boldsymbol{k}_t - \mathrm{i}\theta v_0\,\hat{z} \tag{11.32b}$$

Here $\theta = 1$ for $z > z'$ and $\theta = -1$ for $z < z'$.

Following Eqs. (11.30) and (11.31), the transformed fields can be written as

$$\boldsymbol{E}(\boldsymbol{k}_t, z) = \boldsymbol{E}_t + \hat{z}\, E_z \quad \boldsymbol{H}(\boldsymbol{k}_t, z) = \boldsymbol{H}_t + \hat{z}\, H_z \tag{11.33}$$

where

$$\boldsymbol{E}_t = \mathrm{i}(\omega\varepsilon_0)^{-1} B(k_t, z)[\nu_0 \theta J_z \boldsymbol{w} - (\boldsymbol{w}\boldsymbol{w} + k_0^2 \hat{I})\boldsymbol{J}_t] \tag{11.34a}$$

$$E_z = \mathrm{i}(\omega\varepsilon_0)^{-1} B(k_t, z)(\nu_0 \theta \boldsymbol{w}\boldsymbol{J}_t - k_t^2 J_z) \tag{11.34b}$$

$$\boldsymbol{H}_t = B(k_t, z)[\nu_0 \theta(\hat{x}\,\hat{y} - \hat{y}\,\hat{x})\boldsymbol{J}_t + J_z \boldsymbol{g}] \tag{11.34c}$$

$$H_z = B(k_t, z)\boldsymbol{g}\boldsymbol{J}_t \tag{11.34d}$$

$$B(k_t, z) = (2\nu_0)^{-1} \exp(-\nu_0 |z - z'|) \tag{11.34e}$$

$$\boldsymbol{J}_t = (2\pi)^{-1} \boldsymbol{P}_t \exp(-\boldsymbol{w}\boldsymbol{r}') \quad J_z = (2\pi)^{-1} P_z \exp(-\boldsymbol{w}\boldsymbol{r}') \tag{11.34f}$$

Here $\boldsymbol{w}$ and $\boldsymbol{g}$ are given in Eqs. (11.19) and (11.24), respectively. The branch cuts for the square root $\nu_0 = \sqrt{k_t^2 - k_0^2}$ should be defined so that the outgoing wave conditions are always satisfied. This subject is treated in detail in Section 11.4.

Finally, let us consider the induced fields in the region 2. Following a procedure similar to that for the $\boldsymbol{E}_1$ field in the anisotropic substrate and incorporating the outgoing wave conditions, it is possible to express the transformed field components as follows:

$$\boldsymbol{E}_{2t}(\boldsymbol{k}_t, z) = \frac{1}{\mathrm{i}\omega\varepsilon_0} \begin{pmatrix} k_y & -\nu_0 k_x \\ -k_x & -\nu_0 k_y \end{pmatrix} \begin{pmatrix} C \\ D \end{pmatrix} \mathrm{e}^{-\nu_0 z} \quad E_{2z}(\boldsymbol{k}_t, z) = \frac{k_t^2}{\omega\varepsilon_0} D\mathrm{e}^{-\nu_0 z} \tag{11.35a}$$

$$\boldsymbol{H}_{2t}(\boldsymbol{k}_t, z) = -\begin{pmatrix} \nu_0 k_x k_0^{-2} & -k_y \\ \nu_0 k_y k_0^{-2} & k_x \end{pmatrix} \begin{pmatrix} C \\ D \end{pmatrix} \mathrm{e}^{-\nu_0 z}, \; H_{2z}(\boldsymbol{k}_t, z) = \mathrm{i}\left(\frac{k_t}{k_0}\right)^2 C\mathrm{e}^{-\nu_0 z} \tag{11.35b}$$

It is interesting to notice that in region 2, E_{2z} and H_{2z} are independent (TM and TE waves) while in region 1, because of the tensor nature of $\boldsymbol{\mu}$, the E_{1z} component is a linear combination of the extraordinary (h_{1z}) and ordinary (h_{2z}) magnetic waves.

Up to now, field expressions are derived that satisfy the Maxwell equations, in terms of the unknown coefficients C, D [Eqs. (11.35)] and C_{jz}, $D_{j,z}$ [Eq. (11.18)]. In order to determine these unknowns, the boundary conditions at the interface planes $z = 0$ and $z = -d$ should be employed. To this end, the continuity of the tangential fields for $z = 0$ requires that

$$\boldsymbol{E}_{1t}(\boldsymbol{k}_t) = \boldsymbol{E}_t(\boldsymbol{k}_t) + \boldsymbol{E}_{2t}(\boldsymbol{k}_t) \quad \boldsymbol{H}_{1t}(\boldsymbol{k}_t) = \boldsymbol{H}_t(\boldsymbol{k}_t) + \boldsymbol{H}_{2t}(\boldsymbol{k}_t) \tag{11.36}$$

while at the perfect conductor surface at $z = -d$,

$$\boldsymbol{E}_{1t}(\boldsymbol{k}_t, z = -d) = 0 \tag{11.37}$$

Each of these vector equations can be reduced to two equivalent scalar equations by using a two-dimensional coordinate system in the xy-plane. To this end, the orthogonal system defined by the vectors $\boldsymbol{w}$ and $\mathbf{g}$ ($\boldsymbol{w}\mathbf{g} = 0$) is employed. According to this, the projections of $\boldsymbol{E}_{jt}(\boldsymbol{k}_t, z)$, $\boldsymbol{H}_{jt}(\boldsymbol{k}_t, z)$, $j = 1, 2$ and $\boldsymbol{E}_t$, $\boldsymbol{H}_t$ [see Eqs. (11.34a) and (11.34c)] onto the $\boldsymbol{w}$ and $\mathbf{g}$ vectors satisfy the same equations as Eqs. (11.36) and (11.37). Following the definitions of Eqs. (11.17), (11.20) and (11.22), one can easily prove the identities

$$\boldsymbol{w}\,\hat{M}^{-1}(\tau_j)\boldsymbol{w} = (k_t^2 - k_1^2)\tau_j^{-2} \quad \mathbf{g}\,\hat{M}^{-1}(\tau_j)\boldsymbol{w} = \mathrm{i}\mu_2 k_1^2 k_t^2 \Delta_j^{-1} \tag{11.38a}$$

$$\tau_j \boldsymbol{w}\,\hat{N}(\tau_j)\,\hat{M}^{-1}(\tau_j)\boldsymbol{w} = \mathrm{i}\mu_2 k_1^2 k_t^2 \tau_j^2 \Delta_j^{-1} \quad \tau_j \mathbf{g}\,\hat{N}(\tau_j)\,\hat{M}^{-1}(\tau_j)\boldsymbol{w} = k_1^2 - k_t^2 \tag{11.38b}$$

where

$$\Delta_j = k_t^2[\mu_1\tau_j^2 + (\mu_1^2 - \mu_2^2)k_1^2 - \mu_1 k_t^2] \quad j = 1, 2 \tag{11.39}$$

Taking the inner products of Eqs. (11.36) and (11.37) with $\boldsymbol{w}$ and $\mathbf{g}$, and then using the identities of Eqs. (11.38a) and (11.38b), when $z' > 0$ (source point in the region 2), a 6×6 matrix equation is obtained in the form

$$\begin{pmatrix} R_{11} & 0 & R_{13} \\ R_{21} & R_{22} & 0 \\ 0 & R_{32} & R_{33} \end{pmatrix} \begin{pmatrix} \vec{x}_1 \\ \vec{x}_2 \\ \vec{x}_3 \end{pmatrix} = \begin{pmatrix} \vec{y}_1 \\ 0 \\ \vec{y}_3 \end{pmatrix} \tag{11.40}$$

where the 2×2 submatrices R_{ij} are given by

$$R_{11} = -\varepsilon_0 k_1^2 \begin{pmatrix} \gamma_1 & \gamma_2 \\ 1 & 1 \end{pmatrix}$$

$$R_{13} = \varepsilon_0 \varepsilon k_t^2 \begin{pmatrix} 0 & -\mathrm{i}\nu_0 \\ 1 & 0 \end{pmatrix}$$

$$R_{21} = \begin{pmatrix} \tau_1^2\Delta_1^{-1}\cosh(\tau_1 d) & \tau_2^2\Delta_2^{-1}\cosh(\tau_2 d) \\ \cosh(\tau_1 d) & \cosh(\tau_2 d) \end{pmatrix}$$

$$R_{22} = -\begin{pmatrix} \tau_1^2\Delta_1^{-1}\sinh(\tau_1 d) & \tau_2^2\Delta_2^{-1}\sinh(\tau_2 d) \\ \sinh(\tau_1 d) & \sinh(\tau_2 d) \end{pmatrix}$$

$$R_{32} = -\begin{pmatrix} \gamma_1 k_1^2 \tau_1^{-1} & \gamma_2 k_1^2 \tau_2^{-1} \\ (k_t^2 - k_1^2)\tau_1^{-1} & (k_t^2 - k_1^2)\tau_2^{-1} \end{pmatrix} \tag{11.40b}$$

$$R_{33} = \mathrm{i}k_t^2 \begin{pmatrix} 0 & 1 \\ -k_0^{-2}\nu_0 & 0 \end{pmatrix} \gamma_j \equiv \mathrm{i}\mu_2 k_t^2 \tau_j^2 \Delta_j^{-1}\; j = 1, 2 \tag{11.40c}$$

The unknown and right-hand-side terms are defined as

$$\underset{1}{\vec{x}} = \begin{pmatrix} C_{1z} \\ C_{2z} \end{pmatrix} \quad \underset{2}{\vec{x}} = \begin{pmatrix} D_{1z} \\ D_{2z} \end{pmatrix} \quad \underset{3}{\vec{x}} = \begin{pmatrix} C \\ D \end{pmatrix} \tag{11.41a}$$

$$\underset{1}{\vec{y}} = \varepsilon_0 \varepsilon \nu_0^{-1} \underset{3}{\vec{y}} \quad \underset{3}{\vec{y}} = \frac{1}{2} \mathrm{e}^{-\nu_0 z} \hat{T} \boldsymbol{J}(\boldsymbol{k}_t) \tag{11.41b}$$

where

$$\boldsymbol{J}(\boldsymbol{k}_t) = \boldsymbol{J}_t(\boldsymbol{k}_t) + \hat{z} J_z(\boldsymbol{k}_t) = (2\pi)^{-1} \boldsymbol{p} \exp(\mathrm{i}\boldsymbol{k}_t \boldsymbol{r}') \quad \hat{T} \boldsymbol{J}(\boldsymbol{k}_t) = \begin{pmatrix} a \\ b \end{pmatrix} \tag{11.41c}$$

$$a = -\nu_0^2 (\boldsymbol{w} + \nu_0^{-1} k_t^2 \hat{z}) \boldsymbol{J}(\boldsymbol{k}_t) \quad b = k_0^2 \boldsymbol{g} \boldsymbol{J}(\boldsymbol{k}_t) \tag{11.41d}$$

The system of linear equations in Eq. (11.40) can be solved easily following lengthy but straightforward algebra. Assuming that the 6×6 system of Eq. (11.40) is inverted, it is possible to express the unknown coefficients in Eq. (11.41a) in terms of the primary excitation $\boldsymbol{J} = \boldsymbol{p}\delta(\boldsymbol{r} - \boldsymbol{r})$. The final result is obtained by substituting $C_{jz}, D_{jz} (j = 1, 2)$ and C, D into the field expressions of Eqs. (11.17), (11.21)–(11.23). The electric field (in practice the most interesting part of the solution) for region 1 is found to be

$$\boldsymbol{E}_1(\boldsymbol{k}_t) = -\frac{\Delta_1 \Delta_2 k_1^2 \exp(-\nu_0 z)}{\omega \varepsilon_0 \varepsilon k_t^2 \Delta_\perp} \hat{A} \hat{B} \begin{pmatrix} \nu_0^{-1} b \\ -\nu_0^{-2} a \end{pmatrix} \tag{11.42}$$

where

$$\hat{A} = \begin{pmatrix} \alpha_1 (\gamma_1 k_x + k_y) & \alpha_2 (\gamma_2 k_x + k_y) \\ \alpha_1 (\gamma_1 k_y - k_x) + \mathrm{i}\beta_1 \gamma_1 k_t^2 \tau_1^{-1} & \alpha_2 (\gamma_2 k_y - k_x) + \mathrm{i}\beta_2 \gamma_2 k_t^2 \tau_2^{-1} \end{pmatrix} \tag{11.42a}$$

$$\hat{B} = \begin{pmatrix} \gamma_2 \tau_2^{-1} F_2 & G_2 \\ -\gamma_1 \tau_1^{-1} F_1 & -G_1 \end{pmatrix} \tag{11.42b}$$

$$\alpha_j = \sinh[\tau_j (z + d)] \sinh^{-1}(\tau_j d) \quad \beta_j = \cosh[\tau_j (z + d)] \cosh(\tau_j d) \tag{11.42c}$$

$$F_j = k_0^2 \nu_0^{-1} \tau_j + k_1^2 \coth(\tau_j d) \quad G_j = k_0^2 [1 + (k_t^2 - k_1^2)(\nu_0 \tau_j)^{-1} \coth(\tau_j d)] \tag{11.42d}$$

and the common denominator $\Delta_\perp$ is given by

$$\begin{aligned} \Delta_\perp = {} & \mathrm{i}\mu_2 \nu_0^{-2} k_0^4 k_t^4 \{ \nu_0 (\tau_1^2 - \tau_2^2)(k_2^2 - \varepsilon \tau_1 \tau_2 t_1 t_2) - \tau_1 \tau_2 (\tau_1 t_2 - \tau_2 t_1) [(1 \\ & + \varepsilon \mu_1) k_t^2 - 2\mu_1 k_1^2] + (\tau_1 t_1 - \tau_2 t_2) k_2^2 [(1 + \varepsilon) k_t^2 - 2k_1^2] \} \end{aligned} \tag{11.43}$$

where

$$t_j \equiv \coth(\tau_j d) \quad k_2^2 = k_1^2(\mu_1^2 - \mu_2^2) - \mu_1 k_t^2 \tag{11.43a}$$

It is interesting to notice the symmetry of the solution with respect to τ_1 and τ_2 and the standing wave behavior of the solution due to the $\cosh[\tau_j(z+d)]$ and $\sinh[\tau_j(z+d)]$ terms. As the electric field is obtained in the form of an inner product $\boldsymbol{E}_1 = \hat{G}\boldsymbol{J}$, for an arbitrarily oriented dipole $\boldsymbol{J}$, the dyadic Green's function is also given by Eq. (11.42). The field outside the slab can be derived directly from Eq. (11.42) using the boundary conditions at the $z = 0$ plane interface. Indeed, observing from Eqs. (11.35) that

$$\boldsymbol{E}_{2t}(\boldsymbol{k}_t, z) = \boldsymbol{E}_{2t}(\boldsymbol{k}_t, z=0)\mathrm{e}^{-\nu_0 z} \quad E_{2z}(\boldsymbol{k}_t, z) = E_{2z}(\boldsymbol{k}_t, z=0)\mathrm{e}^{-\nu_0 z} \tag{11.44}$$

for $z > 0$ and since

$$\boldsymbol{E}_t(\boldsymbol{k}_t) + \boldsymbol{E}_{2t}(\boldsymbol{k}_t) = \boldsymbol{E}_{1t}(\boldsymbol{k}_t) \quad E_z(\boldsymbol{k}_t) + E_{2z}(\boldsymbol{k}_t) = \varepsilon E_{1z}(\boldsymbol{k}_t) \tag{11.45}$$

at $z = 0$, the total field for the region 2 can be written as

$$\begin{aligned}\boldsymbol{E}(\boldsymbol{k}_t, z) + \boldsymbol{E}_2(\boldsymbol{k}_t, z) = {} & \boldsymbol{E}(\boldsymbol{k}_t, z) + \mathrm{e}^{-\nu_0 z}\hat{z}\,[\varepsilon E_{1z}(\boldsymbol{k}_t, z=0) - E_z(\boldsymbol{k}_t, z \\ & = 0)] + e^{-\nu_0 z}[\boldsymbol{E}_{1t}(\boldsymbol{k}_t, z=0) - \boldsymbol{E}_t(\boldsymbol{k}_t, z=0)]\end{aligned} \tag{11.46}$$

11.3 Derivation of Green's Function for Parallel Magnetization

When the magnetizing field is parallel to the gyromagnetic slab, it can be assumed, without loss of generality, that the static field is directed along the y-axis. The magnetic permeability tensor in this case takes the form given by Eq. (11.2). A procedure similar to that for the perpendicular magnetization is employed to treat this case also. Now, however, $\boldsymbol{H}_0$ does not coincide any more with the geometrical symmetry z-axis, and additional complexities arise in the analysis. The S_1 and S_2 parameters in Eq. (11.10) are given by Eqs. (11.13) and (11.14), and Eqs. (11.15), (11.16a) and (11.16b) hold also. Again, the tangential magnetic field is expressed in terms of the normal components

$$\boldsymbol{H}_{1t}(\boldsymbol{k}_t, z) = -\sum_{j=1}^{2} \hat{M}^{-1}(\tau_j)[\boldsymbol{w}\tau_j h_{jz}(z) - \mathrm{i}\mu_2 k_1^2 \hat{x}\, h_{jz}^*(z)] \tag{11.47}$$

where the h_{jz} components are given by Eq. (11.18), and

$$\hat{M}(\tau) = (k_y^2 - \tau^2 - \mu_1 k_1^2)\,\hat{x}\,\hat{x} - k_x k_y(\hat{x}\,\hat{y} + \hat{y}\,\hat{x}) + (k_x^2 - \tau^2 - k_1^2)\,\hat{y}\,\hat{y} \tag{11.48}$$

In a similar manner, the electric field tangential components are found to be

$$\boldsymbol{E}_{1t}(\boldsymbol{k}_t, z) = (\mathrm{i}\omega\varepsilon_0\varepsilon)^{-1}\sum_{j=1}^{2}[(\mathbf{g} - \tau_j \hat{N}\hat{M}^{-1}\mathbf{w})h^*_{jz}(z) + \mathrm{i}\mu_2 k_1^2 \hat{N}\hat{M}^{-1}\hat{x}h_{jz}(z)] \tag{11.49}$$

where $\hat{N}(\tau_j)$ is given by Eq. (11.25). The expression for $E_{1z}(\boldsymbol{k}_t, z)$ coincides with that given in Eq. (11.23).

For the field in the region $z > 0$, the same expressions as those of the previous case hold [Eqs. (11.26a)–(11.35b)]. The boundary conditions at $z = 0$ and $z = -d$ are also the same as the $\boldsymbol{H}_0 = H_0\hat{z}$ case [Eqs. (11.36) and (11.37)].

Substituting the field expressions given in Eqs. (11.18), (11.23), (11.47) and (11.49) into the boundary conditions and again using the vectors $\mathbf{w}$ and $\mathbf{g}$ as a basis set, one is able, after lengthy algebraic manipulations, to evaluate the unknown coefficients $C_{jz}, D_{jz,}(j = 1, 2), C$ and D from the following system of equations:

$$\begin{pmatrix} S_{11} & S_{12} & R_{13} \\ S_{21} & S_{22} & 0 \\ S_{31} & S_{32} & R_{33} \end{pmatrix}\begin{pmatrix} \vec{x}_1 \\ \vec{x}_2 \\ \vec{x}_3 \end{pmatrix} = \begin{pmatrix} \vec{y}_1 \\ 0 \\ \vec{y}_3 \end{pmatrix} \tag{11.50}$$

in which the right-hand side is the same as that in Eq. (11.40), since the primary excitation is identical in both cases. The 2×2 submatrices R_{13} and R_{33} are also the same as those of Eq. (11.40). The expressions for $S_{ij}, i = 1, 2, 3,\ j = 1, 2$ are given by

$$S_{11} = \varepsilon_0 k_1^2\begin{pmatrix} \chi_1\Delta_1^{-1} & \chi_2\Delta_2^{-1} \\ \rho_1\Delta_1^{-1} & \rho_2\Delta_2^{-1} \end{pmatrix} \qquad S_{12} = -\varepsilon_0 k_1^2\begin{pmatrix} \varphi_1\Delta_1^{-1} & \varphi_2\Delta_2^{-1} \\ \psi_1\Delta_1^{-1} & \psi_2\Delta_2^{-1} \end{pmatrix} \tag{11.51a}$$

$$\begin{gathered} S_{21} = k_1^2\begin{pmatrix} (\chi_1 C_1 + \varphi_1 S_1)\Delta_1^{-1} & (\chi_2 C_2 + \varphi_2 S_2)\Delta_2^{-1} \\ (\rho_1 C_1 + \psi_1 S_1)\Delta_1^{-1} & (\rho_2 C_2 + \psi_2 S_2)\Delta_2^{-1} \end{pmatrix} \\ S_{31} = k_1^2\begin{pmatrix} -\varphi_1\tau_1^{-1}\Delta_1^{-1} & -\varphi_2\tau_2^{-1}\Delta_2^{-1} \\ \psi_1\tau_1^{-1}\Delta_1^{-1} & \psi_2\tau_2^{-1}\Delta_2^{-1} \end{pmatrix} \end{gathered} \tag{11.51b}$$

$$\begin{gathered} S_{22} = -k_1^2\begin{pmatrix} (\chi_1 S_1 + \varphi_1 C_1)\Delta_1^{-1} & (\chi_2 S_2 + \varphi_2 C_2)\Delta_2^{-1} \\ (\rho_1 S_1 + \psi_1 C_1)\Delta_1^{-1} & (\rho_2 S_2 + \psi_2 C_2)\Delta_2^{-1} \end{pmatrix} \\ S_{32} = k_1^2\begin{pmatrix} \chi_1\tau_1^{-1}\Delta_1^{-1} & \chi_2\tau_2^{-1}\Delta_2^{-1} \\ \dfrac{r_1\tau_1}{k_1^2\Delta_1} & \dfrac{r_2\tau_2}{k_1^2\Delta_2} \end{pmatrix} \end{gathered} \tag{11.51c}$$

where

$$\chi_j = (\mu_1 - 1)k_x k_y \tau_j^2 \qquad \varphi_j = \mu_2 k_y \tau_j (k_1^2 + \tau_j^2) \qquad \psi_j = \mu_2 k_x \tau_j A_j \tag{11.52a}$$

$$A_j = k_t^2 - k_1^2 - \tau_j^2 \quad r_j = k_t^2 A_j - \varsigma \quad \varsigma = (\mu_1 - 1)k_y^2 k_1^2 \quad C_j = \cosh(\tau_j d) \tag{11.52b}$$

$$\rho_j = (\mu_1 - 1)k_x^2 A_j + r_j \quad \Delta_j = \varsigma - (\mu_1 k_1^2 + \tau_j^2)A_j \quad S_j = \sinh(\tau_j d) \tag{11.52c}$$

A comparison of Eq. (11.50) with the corresponding equation (11.40) shows that when $\boldsymbol{H}_0 = \hat{y} H_0$, the solution is considerably involved algebraically. Finally, substituting the solution of Eq. (11.50) into the field expressions (11.23) and (11.49) and following a long sequence of algebraic operations, the field in region 1 is found to be

$$\boldsymbol{E}_1(\boldsymbol{k}_t, z) = -\frac{\exp(-\nu_0 z')}{\Delta_{||} k_0^2 \omega \varepsilon_0 \varepsilon} [\hat{Q}(\tau_1, \tau_2) - \hat{Q}(\tau_2, \tau_1)] \boldsymbol{J}(\boldsymbol{k}_t) \tag{11.53}$$

where the common denominator is

$$\Delta_{||} = L_2(\tau_1, \tau_2) L_1(\tau_2, \tau_1) - L_2(\tau_2, \tau_1) L_1(\tau_1, \tau_2) \tag{11.54}$$

$$\hat{Q}(\tau_1, \tau_2) = \boldsymbol{v}(\tau_2, \tau_1) \boldsymbol{u}(\tau_1, \tau_2) \tag{11.55}$$

$$\boldsymbol{u}(\tau_1, \tau_2) = [\nu_0(k_x \hat{x} + k_y \hat{y}) + ik_t^2 \hat{z}] L_2(\tau_1, \tau_2) + (k_y \hat{x} - k_x \hat{y}) \nu_0^{-1} k_0^2 L_1(\tau_1, \tau_2) \tag{11.55a}$$

$$\boldsymbol{v}(\tau_1, \tau_2) = \nu_1(\tau_1, \tau_2) \boldsymbol{z}_1 + \nu_2(\tau_1, \tau_2) \boldsymbol{z}_2 + \nu_3(\tau_1, \tau_2) \boldsymbol{z}_3 \tag{11.55b}$$

$$\nu_1(\tau_1, \tau_2) = (\mu_1 - 1)k_x C_2(\rho_2 \tau_1^2 - \rho_1 \tau_2^2) + \mu_2 \tau_2 S_1[(\mu_1 - 1)k_x^2 \tau_1^2 A_2 - \rho_1(k_t^2 - A_2)] \tag{11.55c}$$

$$\nu_2(\tau_1, \tau_2) = \mu_2 \tau_1 [\mu_2 k_x k_t^2 \tau_2 S_1 (\tau_1^2 - \tau_2^2) - C_2\{(\mu_1 - 1)k_x^2 \tau_2^2 A_1 - \rho_2(k_t^2 - A_1)\}] \tag{11.55d}$$

$$\nu_3(\tau_1, \tau_2) = \tau_1[(\mu_1 - 1)k_x^2 \tau_1 A_1 - \mu_2 \rho_1 (k_t^2 - A_1)] \tag{11.55e}$$

$$\boldsymbol{z}_1 = \boldsymbol{P}(\tau_1) \sinh[\tau_1(z + d)] + \boldsymbol{Q}(\tau_1) \cosh[\tau_1(z + d)] \tag{11.55f}$$

$$\boldsymbol{z}_2 = \boldsymbol{P}(\tau_1) \cosh[\tau_1(z + d)] + \boldsymbol{Q}(\tau_1) \sinh[\tau_1(z + d)] \tag{11.55g}$$

$$\boldsymbol{z}_3 = \boldsymbol{P}(\tau_2) \cosh(\tau_2 z)] + \boldsymbol{Q}(\tau_2) \sinh(\tau_2 z) \tag{11.55h}$$

$$\boldsymbol{P}(\tau) = -ik_y k_1^2[\tau^2 - k_y^2 + \mu_1(k_1^2 - k_x^2)]\hat{x} + ik_x k_1^2[\mu(k_1^2 - k_x^2 + \tau^2) - k_y^2]\hat{y} + \mu_2 k_y k_1^2 (k_1^2 + \tau^2)\hat{z} \tag{11.55i}$$

$$Q(\tau) = k_1^2\tau[-i\mu_2 k_x k_y \hat{x} + i\mu_2(k_x^2 - k_1^2 - \tau^2)\hat{y} - (\mu_1 - 1)k_x k_y \hat{z}] \tag{11.55j}$$

$$\begin{aligned} L_1(\tau_1, \tau_2) = k_y\{&\tau_1[-\mu_2 B_2 D_1 + C_1 C_2(\mu_2 B_1 D_2 - G_1 G_2) + C_1 S_2 \mu_2 \tau_2 (G_1 F_1 - B_1 G_3)] \\ &+ S_1 C_2 (B_1 G_2 - \mu_2 \tau_1^2 D_2 G_1) + S_1 S_2 \mu_2 \tau_2 (\tau_1^2 G_1 G_3 - B_1 F_1)\} \end{aligned} \tag{11.56}$$

$$B_j = (\mu_1 - 1)k_x \tau_j^2 - \varepsilon v_0 \mu_2 (k_t^2 - A_j) \quad D_j = (\mu_1 - 1)k_x^2 \tau_j^2 A_1 - (k_t^2 - A_1)\rho_j j = 1, 2 \tag{11.56a}$$

$$F_j = (\mu_1 - 1)k_x^2 \tau_j^2 A_2 - (k_t^2 - A_2)\rho_j \quad G_1 = \mu_2(k_t^2 - A_1) - \varepsilon v_0 (\mu_1 - 1)k_x \tag{11.56b}$$

$$G_2 = (\mu_1 - 1)k_x(\rho_1 \tau_2^2 - \rho_2 \tau_1^2) \quad G_3 = \mu_2 k_x k_t^2 (\tau_1^2 - \tau_2^2) \tag{11.56c}$$

$$\begin{aligned} L_2(\tau_1, \tau_2) = &\tau_1[-\mu_2 M_2 D_1 + C_1 C_2(\mu_2 M_1 D_2 - G_0 G_2) + C_1 S_2 \mu_2 \tau_2 (G_0 F_1 - M_1 G_3)] \\ &+ S_1 C_2 (M_1 G_2 - \mu_2 \tau_1^2 D_2 G_0) + S_1 S_2 \mu_2 \tau_2 (\tau_1^2 G_0 G_3 - M_1 F_1)] \end{aligned} \tag{11.57}$$

$$M_j = \rho_j + \mu_2 k_x k_1^2 A_j v_0^{-1} \qquad G_0 = \mu_2 k_x A_1 - r_1 v_0^{-1} \tag{11.57a}$$

The field in region 2 can be computed using Eq. (11.46), since the same boundary conditions as those in Eq. (11.45) hold.

11.4 Far-Field Behavior

Following the results of previous sections, the electric field in region 2 for both magnetization cases can be computed from a Fourier transformation:

$$\mathbf{E}_2(\mathbf{r}) = \frac{1}{2\pi} \iint d\mathbf{k}_t \mathbf{F}(\mathbf{k}_t) \exp[-i\mathbf{k}_t(\mathbf{r} - \mathbf{r}') - v_0 z] \tag{11.58}$$

where $\mathbf{F}(\mathbf{k}_t)\exp(i\mathbf{k}_t\mathbf{r}') = \mathbf{E}(\mathbf{k}_t, 0) + \mathbf{E}_2(\mathbf{k}_t, 0)$ is defined in Eqs. (11.44)–(11.46) and the primary excitation is assumed to be located at the $z' = 0$ plane.

Inspection of Eqs. (11.42) and (11.53) shows that $\mathbf{F}(\mathbf{k}_t)$ is an analytic function apart from the singularities that arise from zeros of the denominator Δ. Indeed, recalling [144] the well-known behavior of the $\mathbf{F}(\mathbf{k}_t)$ function for isotropic substrates, it can be asserted that the roots of $\Delta(k_x, k_y) = 0$ correspond to the surface waves supported from the grounded gyromagnetic slab. This point will be analyzed later on in detail. After the singularities of $\mathbf{F}(\mathbf{k}_t)$ are found, several integration techniques can be used to evaluate the Fourier transform of Eq. (11.58).

Instead of Cartesian coordinates $\boldsymbol{k}_t = k_x\hat{x} + k_y\hat{y}$ it is possible to use the polar coordinates $\boldsymbol{k}_t = \lambda(\cos\chi\,\hat{x} + \sin\chi\,\hat{y})$, $\lambda \equiv k_t$. Then, Eq. (11.58) can be written in the form

$$\boldsymbol{E})_2(\boldsymbol{r}) = \int_0^{2\pi} \mathrm{d}x \int_c \lambda \boldsymbol{F}_1(\lambda, \chi)\exp[-\boldsymbol{u}(\boldsymbol{r} - \boldsymbol{r}')\mathrm{d}\lambda \tag{11.59}$$

where

$$\boldsymbol{u} = \lambda(\cos\chi\hat{x} + \sin\chi\,\hat{y}) - iv_0\,\hat{z} \quad \boldsymbol{F}_1(\lambda, \chi) = \frac{1}{2\pi}\boldsymbol{F}(k_x, k_y) \tag{11.59a}$$

and the contour c is defined in Fig. 11.2, in which the branch cuts for the $v_0 = \sqrt{(\lambda - k_0)(\lambda + k_0)}$ function are also shown. The choice of these particular branch cuts was determined after requiring that the factor exp(–i$\boldsymbol{ur}$) should satisfy the outgoing wave conditions. For an arbitrary point $\boldsymbol{r}$, the evaluation of Eq. (11.59) is rather involved. The Cauchy theorem can be applied by closing the contour c with the path B ($z > 0$), as shown in Fig. 11.2. The result can then be obtained in terms of a summation over the residues of the singularities, giving rise to the surface-wave terms, and the contour integration on both sides of the branch cut originating at $k = k_0$. Another direct numerical algorithm can also be used to evaluate the integral as described in Ref. [145] for isotropic substrates. If the observation point $\boldsymbol{r}$ is far from the source point $\boldsymbol{r}'$, it is possible to evaluate $\boldsymbol{E}_2(\boldsymbol{r})$ approximately using the steepest descent integration technique. To this end the integration variable λ is transformed as follows:

$$\lambda = k_0\sin\beta \tag{11.60}$$

Then, the square root v_0 becomes $v_0 = ik_0\cos\beta$. According to the transformation (11.60), the proper Riemann sheet of the λ-plane (Fig. 11.2) maps onto the shaded areas of Fig. 11.3. Substituting Eq. (11.60) into Eq. (11.59) and with $\boldsymbol{r} = r\,(\cos\varphi\sin\theta\,\hat{x} + \sin\varphi\sin\theta\,\hat{y} + \cos\theta\,\hat{z})$, one obtains

$$\boldsymbol{E}_2(\boldsymbol{r}) = \int_0^{2\pi} \mathrm{d}\chi \int_c \mathrm{d}\beta \boldsymbol{u}(\beta, \chi)\exp(-irk_0\phi) \tag{11.61}$$

where

$$\phi(\beta, \theta, \chi, \varphi) = \sin\theta\sin\beta\cos(\varphi - \chi) + \cos\theta\cos\beta \tag{11.61a}$$

$$\boldsymbol{u}(\beta, \chi) = k_0^2\cos\beta\sin\beta\,\boldsymbol{F}_1(k_0\sin\beta, \chi)\exp(i\boldsymbol{u}\boldsymbol{r}') \tag{11.61b}$$

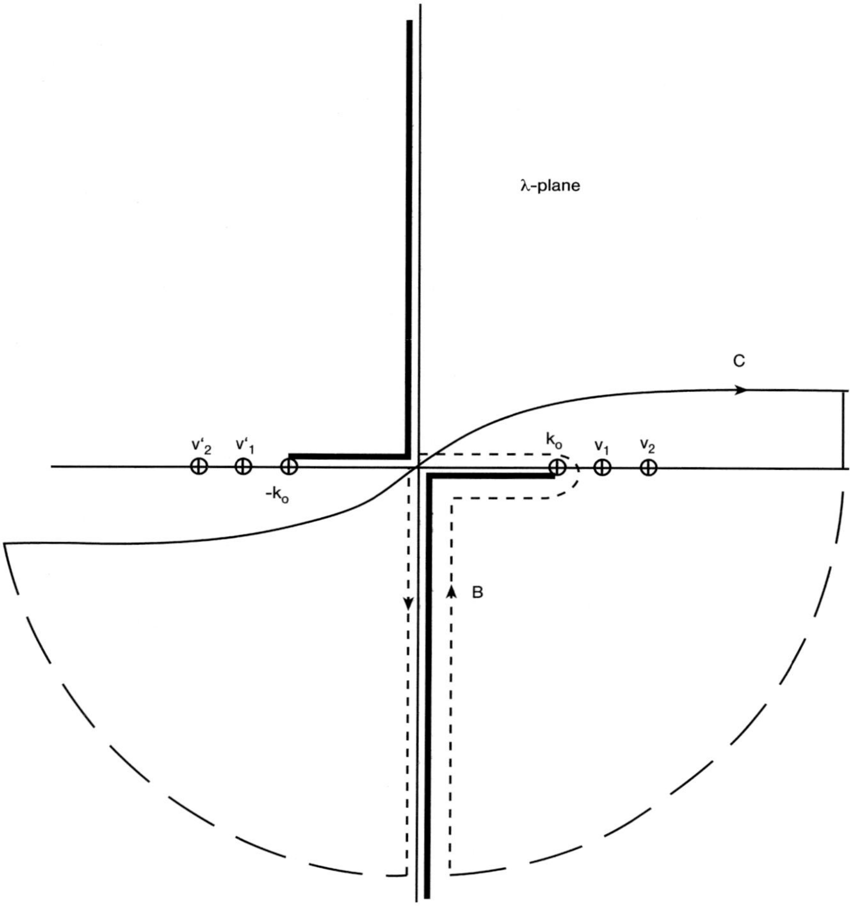

Fig. 11.2 Complex λ plane.

and c′ is the conformal mapping of the contour c of the λ plane into the β plane, as shown in Fig. 11.3. The main contribution to the integral (11.61) as $r \to +\infty$, arises from points at which the phase ϕ is stationary. The saddle point, determined from the equations $\partial\phi/\partial\beta = \partial\phi/\partial\chi = 0$, is found to be at $\beta = \theta$ and $\varphi = \chi$. Consider first the integration over the χ variable. As $r \to +\infty$ the term $\exp[-irk_0 \cos(\varphi - \chi)\sin\theta\sin\beta]$ oscillates rapidly and the integral can be evaluated with the stationary-phase approximation

$$\int_0^{2\pi} d\chi \boldsymbol{u}(\beta,\chi)\exp[-irk_0 \sin\theta\sin\beta\cos(\varphi-\chi) \approx \boldsymbol{u}(\beta,\varphi)e^{-irk_0\sin\theta\sin\beta}$$

$$\times \int_{-\infty}^{+\infty} d\chi e^{i\frac{r}{2}k_0(\varphi-\chi)^2\sin\theta,\sin\beta} \approx \boldsymbol{u}(\beta,\varphi)e^{-irk_0\sin\theta\sin\beta+i\{\pi/4}\sqrt{\frac{2\pi}{k_0 r\sin\theta\sin\beta}} \tag{11.62}$$

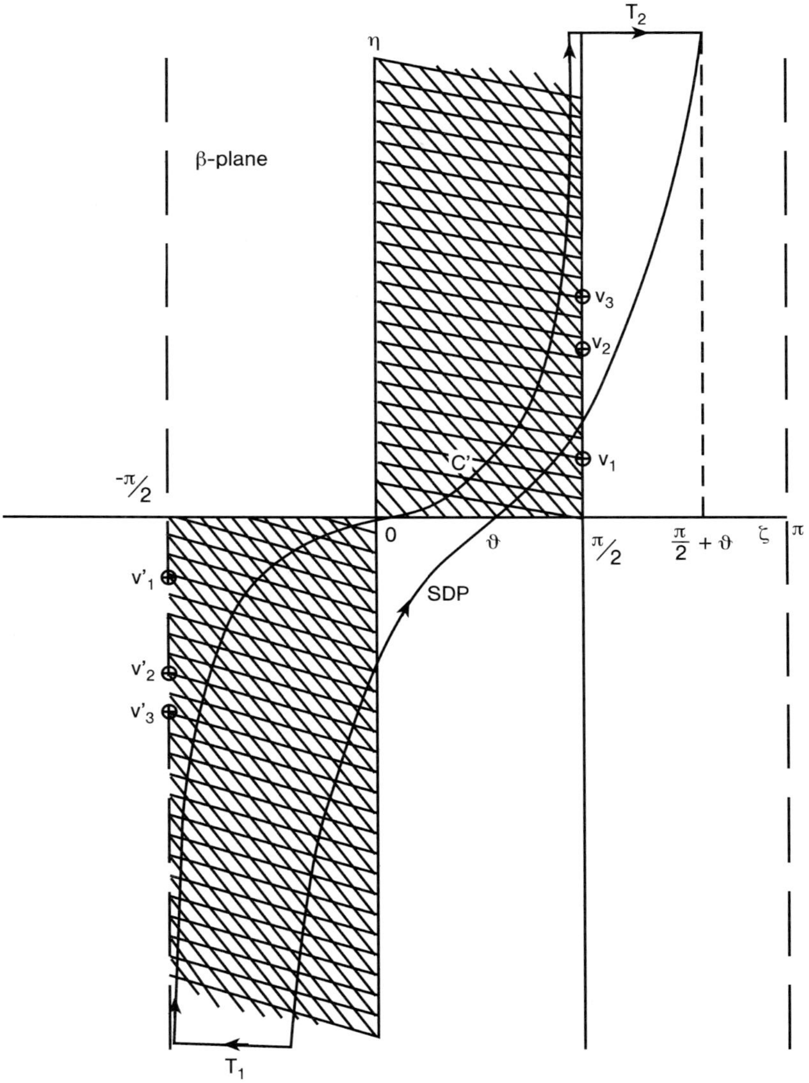

Fig. 11.3 Complex β plane for the steepest descent integration.

So, Eq. (11.61) can be rewritten in the form

$$E_2(r) \approx \int_{C'} d\beta u(\beta, \varphi) \sqrt{\frac{2\pi i}{k_0 r \sin\theta \sin\beta}} \exp[-irk_0 \cos(\theta - \beta)] \tag{11.63}$$

The integration over the variable β can be performed by using the well-known steepest descent approach as follows. As already mentioned, the saddle point is found to be at $\beta = \theta$, and the steepest descent path (SDP) is defined by the equation

$$\boldsymbol{Im}[\mathrm{i}\cos(\theta - \beta) - \mathrm{i}] = 0 \tag{11.64a}$$

Or, substituting $\beta = \xi + in$ by

$$\cos(\theta - \xi)\cosh(n) = 1 \tag{11.64b}$$

The SDP is shown in Fig. 11.3. Its asymptotic values for $n \to \mp \infty$ are $\xi = \theta \pm \pi/2$, respectively. Applying the Cauchy theorem to the closed contour $c'T_1(-SDP)T_2$ we can write the integral as

$$\begin{aligned} \boldsymbol{E}_2(\boldsymbol{r}) = & \left[\int_{SDP} - \int_{T_1} - \int_{T_2} \right] \exp[-\mathrm{i}rk_0\cos(\theta - \beta)]\boldsymbol{u}'(\beta, \varphi)\mathrm{d}\beta \\ & + 2\pi\mathrm{i} \sum_i \operatorname*{Re\,s}_{\lambda \to v_i} [\boldsymbol{u}'(\beta, \varphi]\mathrm{e}^{-\mathrm{i}\rho v,\varphi_e - zv_0(v_i)} \\ & + \sum_j \operatorname*{Re\,s}_{\lambda \to p_j + \mathrm{i}q_j} [\boldsymbol{u}'(\beta, \varphi]\mathrm{e}^{-\mathrm{i}\rho v,\varphi_e - zv_0(p_j + \mathrm{i}q_j)} \end{aligned} \tag{11.65}$$

where

$$\boldsymbol{u}'(\beta, \varphi) = \boldsymbol{u}(\beta, \varphi)\sqrt{\frac{2\pi\mathrm{i}}{k_0 r \sin\theta \sin\beta}} \qquad \rho = \sqrt{x^2 + y^2} \tag{11.66}$$

and the two summations correspond to the contributions of propagating surface waves as well as the leaky or evanescent waves. The propagating surface waves are finite in number and their propagation constants are real since $\lambda = v_i(\varphi)$. In general, they are functions of the azimuthal angle φ, because of the gyromagnetic character of the substrate. However, when $\boldsymbol{H}_0 = H_0\hat{z}$ owing to the axial symmetry, the v_i roots, like those for isotropic substrates, are independent of the angle φ. The singular points $\lambda = p_j + \mathrm{i}q_j$ $(p_j, q_j > 0)$ give rise to the leaky modes whose properties are discussed in some detail in Sect. 11.5. The number of terms contributing to the summations in Eq. (11.65) depends upon the number of residues enclosed by the closed contour $c'T_1(-SDP)T_2$. As $\theta \to \pi/2$, the field is computed for points near the substrate, and all the propagating surface waves are included in Eq. (11.65). The integration along the paths T_1 and T_2 vanishes since $\exp[-\mathrm{i}k_0 r\cos(\theta - \beta)] = \exp[k_0 r\sin(\theta - \xi)\sinh(n)]$ and $\boldsymbol{u}(\beta, \varphi)$ is a regular function. Indeed, on T_1 we have $\sin(\theta - \xi)\sinh(n) < 0$ and the integrand vanishes exponentially as $n \to -\infty$. The same result holds true, also, for the T_2 contour. So, the only contribution to the first term of Eq. (11.65) comes from the SDP. The evanescent and leaky wave terms also vanish as $\rho \rightarrow +\infty$. The final result can be written in the form

$$E_2(r) = u'(\theta, \varphi)\sqrt{\frac{2\pi}{rk_0}}\exp(-irk_0 + i\pi/4) + \text{propagating surface-wave residues} \tag{11.67}$$

where the distance r is assumed sufficiently large so that the surface-wave singularities will not affect the expansion of $u'(\beta, \varphi)$ around the $\beta = \theta$ point. Of course, if necessary, one can apply the well-known procedure to take into account the effect of singularities on the SDP integration [144]. In practice, usually, it is desired to evaluate the field for points far from the substrate surface, when both z and ρ are very large. Then, since $v_0(v_i)z > 0$, the propagating surface wave can also be neglected. The final result is written by substituting Eqs. (11.66) and (11.61b) into Eq. (11.67):

$$E_2(r) \approx ik_0 \frac{e^{-ik_0 r}}{r} \cos\theta F(k_0 \sin\theta\cos\varphi, k_0 \sin\theta\sin\varphi)e^{ik_0 \sin\theta(x'\cos\varphi + y'\sin\varphi)} \tag{11.68}$$

which is valid when both z and ρ approach infinity.

11.5 Numerical Results

The spectrum of propagating surface waves has been investigated numerically using the analytical results of the previous sections. The solution of the corresponding characteristic equation is obtained using the Newton–Raphson algorithm. Since the anisotropy is rather weak ($\mu_2 = 0.28$), the spectrum of the modes is similar to those for isotropic substrates. However, the transverse (TM and TE) character of the modes is destroyed. This is shown in Fig. 11.4, where the dominant-mode field components are compared for an almost isotropic ($\mu_2 = 0.017$) and an anisotropic ($\mu_2 = 0.28$) substrate, with the propagation direction along the x-axis and the static field $H_0 = H_0\hat{z}$. When $\mu_2 = 0.017$, only H_y and E_x, E_z components have significant values, and for all practical purposes the propagating mode is a TM one. For $\mu_2 = 0.28$, the propagating mode has a hybrid character since both H_x and E_x are present. It is also interesting to notice the dependence of the propagation constant on the azimuthal angle φ when $H_0 = H_0\hat{y}$ (parallel magnetization). For strong anisotropies the surface mode pattern deviates completely from the pattern of the isotropic case. As a result of this, the analytical continuation, which is valid for weak anisotropy, cannot be used any more. It can be shown that for $\mu_2 \leq \mu_1$, evanescent surface waves can also be present around the source region. In this case, the spectrum of the waves becomes complicated and will not be treated here.

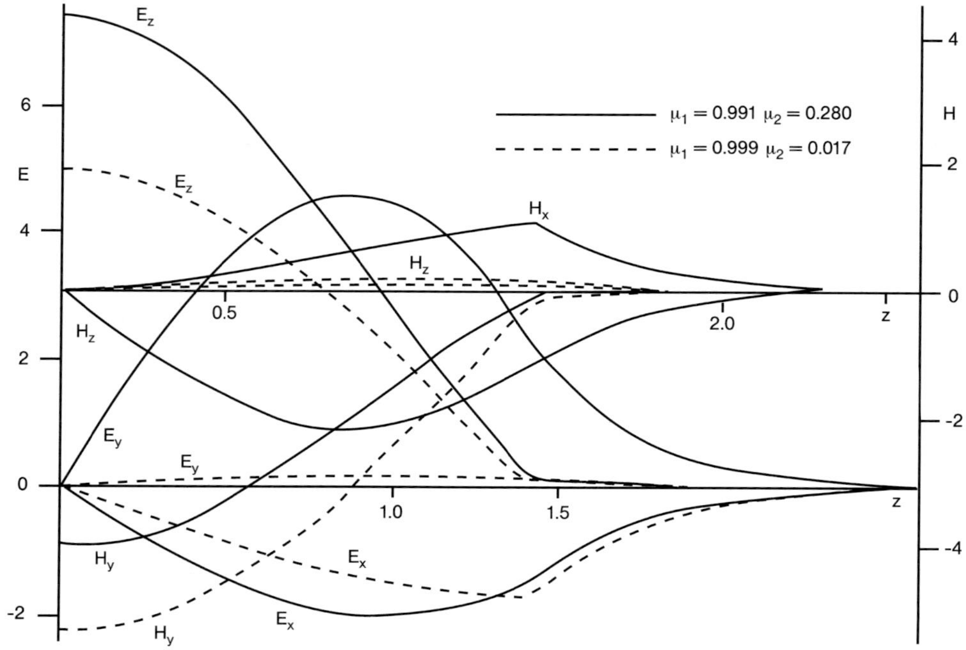

Fig. 11.4 Field distributions for the dominant surface waves propagating along weakly and strongly anisotropic substrates with $H_0 = H_0 \hat{y}$, $k_0 d = 1.36$ and $\varepsilon = 15$.

Numerical results are also obtained for the far field using Eq. (11.68) for both the perpendicular and the parallel magnetization cases. In these computations the impressed current distribution $\boldsymbol{J} = \boldsymbol{p}\delta(\boldsymbol{r} - \boldsymbol{r}')$ is assumed to be constant. In Fig. 11.5a and b, radiation patterns are given for perpendicularly magnetized substrates at 30-GHz frequency. The shapes of the patterns are similar for different μ_1, μ_2 values and are symmetric about the z-axis for a constant φ angle. In Fig. 11.6, results are given for $\boldsymbol{H}_0 = H_0 \hat{y}$ and $\boldsymbol{p} = p_x \hat{x}$. In this case and when $\varphi = 0$, the radiation of E_θ is independent of μ_1, μ_2 for $k_y = 0$. The physical explanation of this simple result is that, with the dipole axis perpendicular to $\boldsymbol{H}_0$, the propagation in the $\varphi = 0$ plane is not affected by the μ_1, μ_2 tensor elements. In the $\varphi = 90°$ plane, the E_θ, E_φ patterns are functions of the anisotropy, and E_θ is a strong function of the μ_1, μ_2 parameters as compared with the E_φ patterns. The E_θ field is due to the anisotropy and, as $\mu_1 \rightarrow 1, \mu_2 \rightarrow 0$, it vanishes. However, for strong anisotropies the magnitude of E_θ is comparable to that of E_φ. The shifting of the main beam with the variation of the μ_1, μ_2 parameters is shown in Fig. 11.7, in which the E_θ, E_φ patterns are given for the $\varphi = 0\,(\pi)$ and $\varphi = \pi/2\,(3\pi/2)$ planes, the dipole orientation being $\boldsymbol{p} = p_y \hat{y}$. It is interesting to notice that for the $\varphi = 0\,(\pi)$ plane, the E_θ field component, that is, in a direction perpendicular to the anisotropy axis, is also zero ($k_y = 0$). The gradual variation from a symmetric pattern to a directed beam is also noticeable as the anisotropy becomes stronger. For inter-

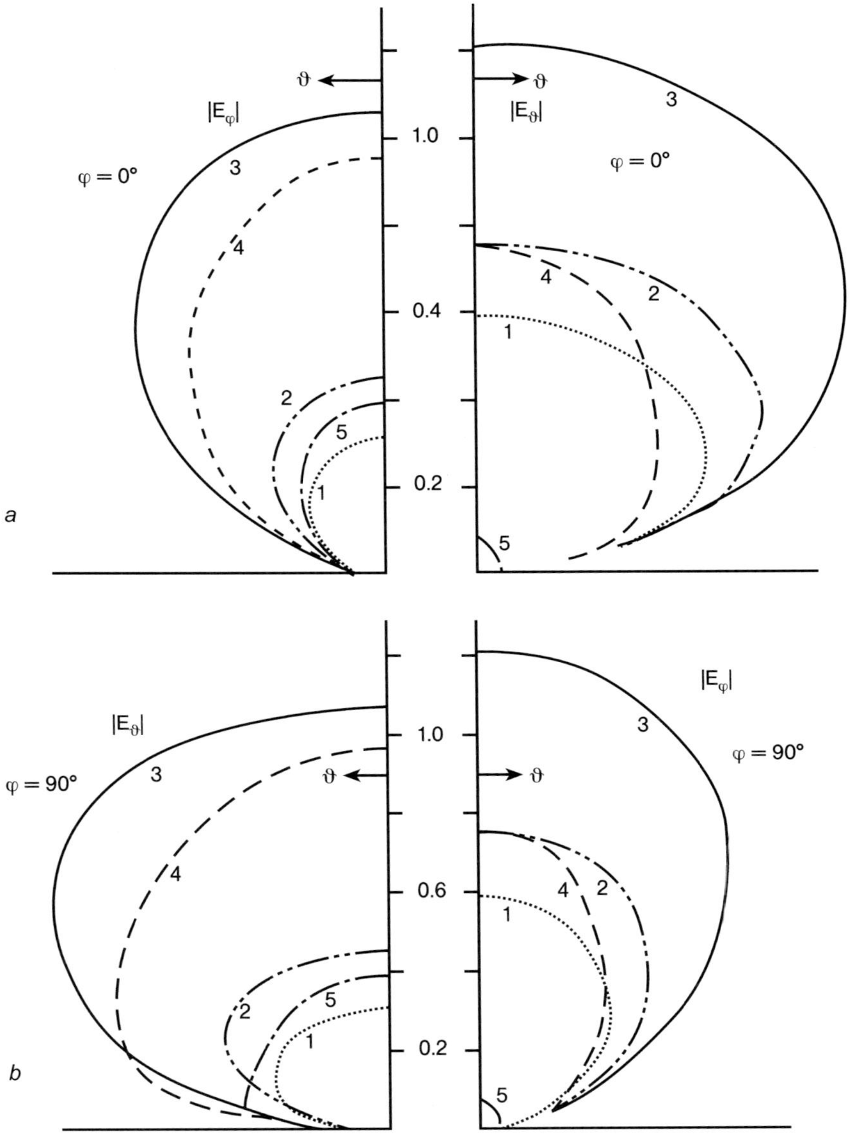

Fig. 11.5 Radiation patterns for a horizontal dipole in the presence of a vertically magnetized gyromagnetic substrate at 30GHz radiation frequency, $d = 1$ mm, $\varepsilon = 15$, and magnetic permeability tensor elements μ_1, μ_2 given as follows: 1.0000, 0.2801 (curve 1); 0.9148, 0.3039 (curve 2); 0.8167, 0.3707 (curve 3); 0.7600, 0.4179 (curve 4); 0.6746, 0.4942 (curve 5).

mediate angles radiation patterns show similar behavior. In Fig. 11.8 the E_θ, E_φ patterns are shown for the $\varphi = 50°$ (230°) plane. The E_θ component, which is mostly affected by the anisotropy, exhibits secondary lobes. In this case, E_φ is also strongly dependent on the μ_1, μ_2 values.

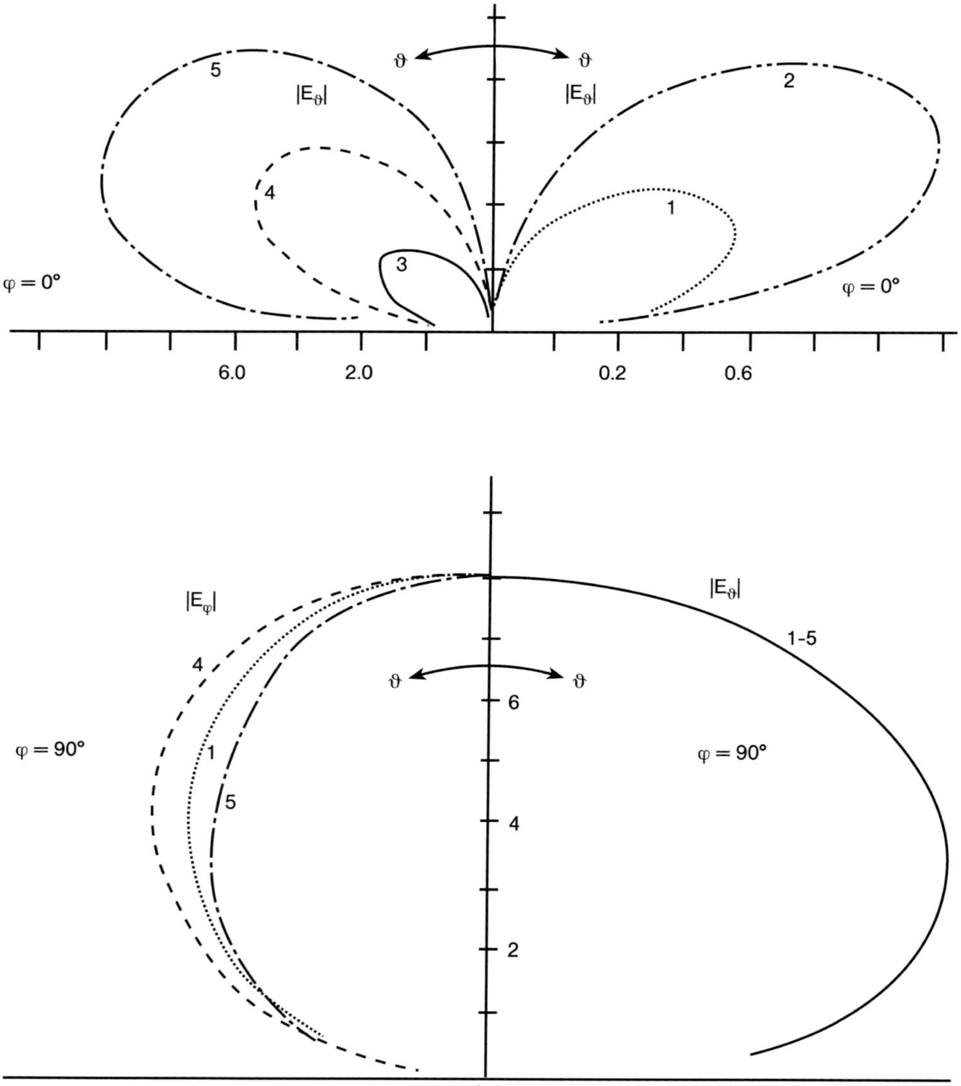

Fig. 11.6 Radiation patterns for $\boldsymbol{H}_0 = H_0\,\hat{y}, \hat{p} = \hat{x}$ and the gyromagnetic substrates defined in Fig. 11.5.

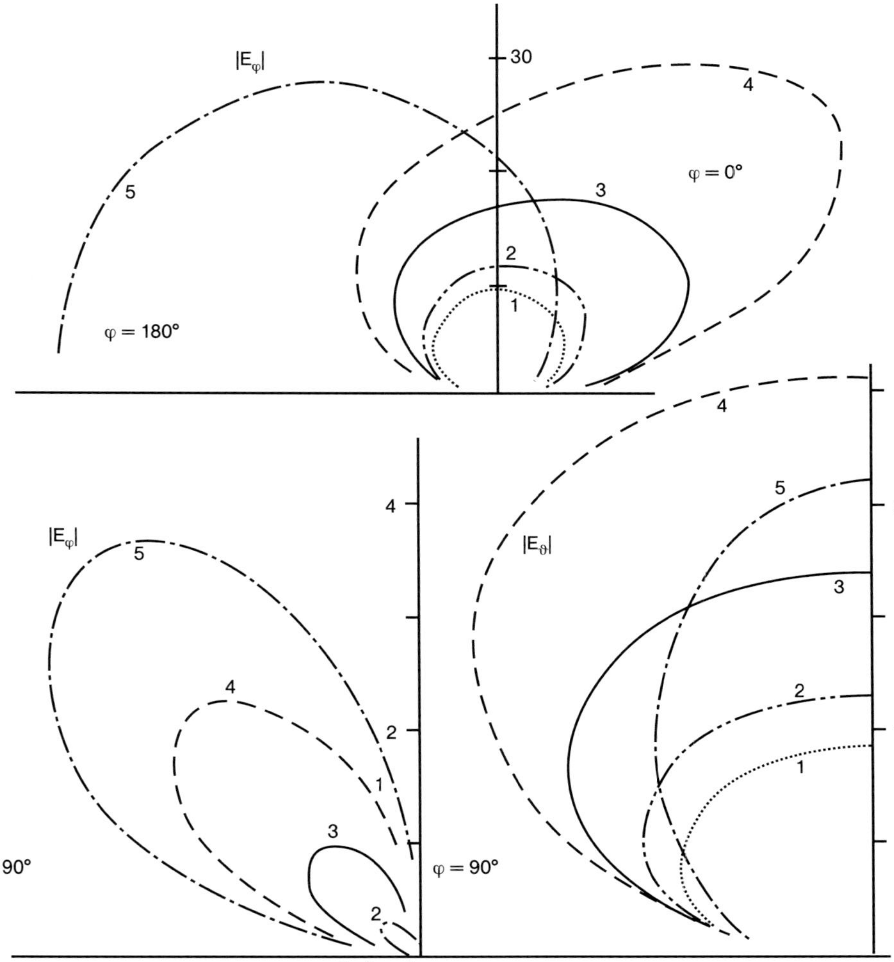

Fig. 11.7 Radiation patterns for $\boldsymbol{H}_0 = H_0\,\hat{y}$, $\hat{p} = \hat{y}$ and the substrates of Fig. 11.5.

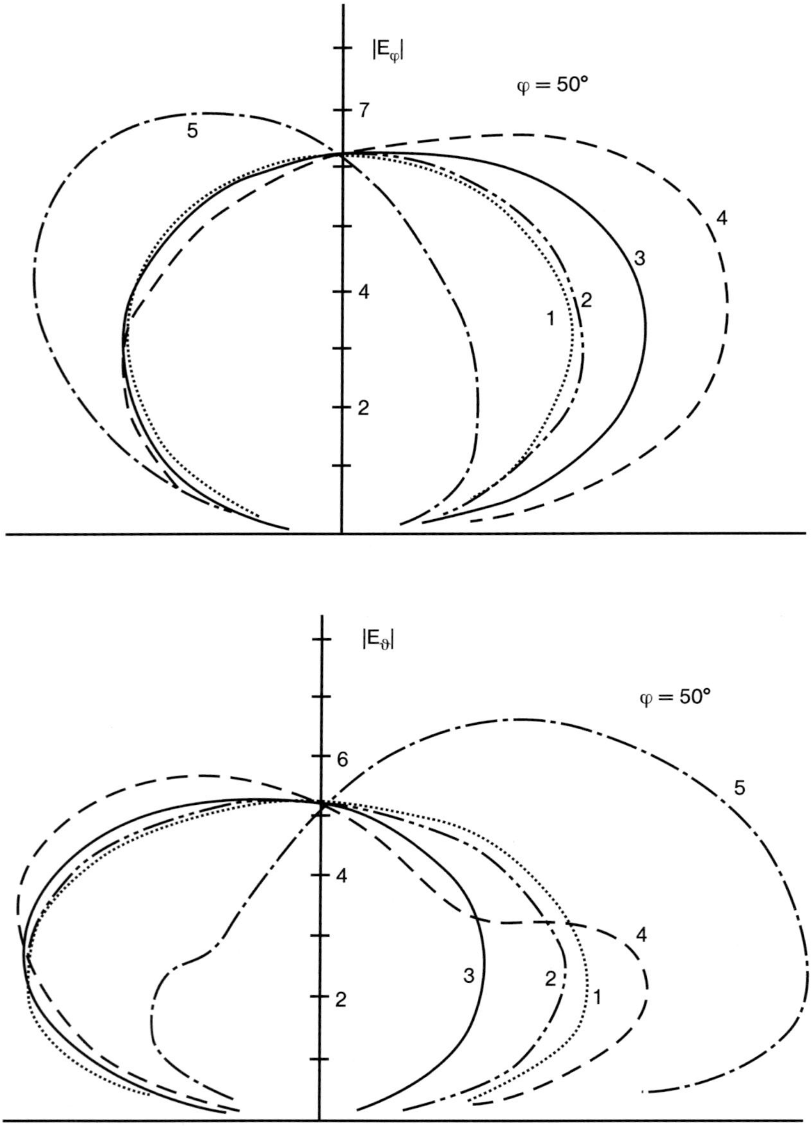

Fig. 11.8 Radiation patterns for the $\varphi = 50°$, $230°$ planes, $\boldsymbol{H}_0 = H_0\,\hat{y}$, $\hat{p} = \hat{x}$ and the substrates of Fig. 11.5.

The field expressions derived in this chapter are directly applicable to several printed microwave circuit (microstrip) problems.

References

1 J. F. Nye, *Physical Properties of Crystals*, Clarendon, Oxford (1957)
2 M. Born and E. Wolf, *Principles of Optics*, Pergamon, London (1959)
3 L. D. Landau and E. M. Lifshitz, *Electrodynamics of Continuous Media*, Pergamon, London (1960)
4 F. I. Fedorov, *Optics of Anisotropic Media*, Izd. Acad. Sci. BSSR, Minsk (1958)
5 V. M. Agranovich and V. L. Ginzburg, *Crystal Optics with Spatial Dispersion and Theory of Excitons*, Springer, New York (1984)
6 A. V. Shubnikov, *Principles of Optical Crystallography*, Izd. Acad. Sci. USSR, Moscow (1958)
7 V. P. Silin and A. A. Rukhadze, *Electrodynamics of Plasma and Plasma-like Mediums*, Atomizdat, Moscow (1961)
8 P. M. Platzman and P. A. Wolff, *Waves and Interactions in Solid State Plasma*, Academic, New York and London (1973)
9 T. Stix, *Waves in Plasmas*, AIP, New York (1992)
10 S. I. Pekar, JETP **33**, 1022 (1957); **34**, 1176 (1958)
11 J. J. Horfield and D. G. Thomas, Phys. Rev. **132**, 563 (1963)
12 G. Agarval, D. Pattanayak and E. Wolf, Phys. Rev. Lett. **27**, 1022 (1971)
13 J. Birman and J. Sein, Phys. Rev. B **6**, 2482 (1972)
14 V. M. Agranovich and V. I. Yudson, Opt. Commun. **7**, 121 (1973)
15 A. A. Maradudin and D. L. Mills, Phys. Rev. B **7**, 2737 (1973)
16 S. I. Pekar and M. I. Strashnikova, JETP **68**, 2047 (1975)
17 V. A. Kiselev, B. S. Rasbirin and I. N. Uraltsev, Phys. Stat. Sol. (b) **72**, 161 (1975)
18 R. H. Tarkhanyan, Sov. Phys. Semicond. **3**, 1451 (1970)
19 L. E. Gurevich and R. H. Tarkhanyan, Phys. Stat. Sol. (b) **58**, 379 (1973)
20 R. H. Tarkhanyan, Phys. Stat. Sol. (b) **64**, K93 (1974)
21 L. E. Gurevich and R. H. Tarkhanyan, Sov. Phys. Semicond. **3**, 47 (1969)
22 L. E. Gurevich and R. H. Tarkhanyan, Sov. Phys. Semicond. **6**, 1482, 1631 (1973)
23 K. Huang, Proc. R. Soc. A **208**, 352 (1951)
24 M. Born and K. Huang, *Dynamical Theory of Crystal Lattices*, Clarendon, Oxford (1954)
25 B. B. Varga, Phys. Rev. **137A**, 1896 (1965)
26 I. Yokota, J. Phys. Soc. Jpn. **16**, 2075 (1961)
27 A. Mooradian and G. Wright, Phys. Rev. Lett. **16**, 999 (1966)
28 L. E. Gurevich and R. H. Tarkhanyan, Sov. Phys. Semicond. **3**, 962 (1970)
29 L. E. Gurevich and R. H. Tarkhanyan, Phys. Stat. Sol. (b) **66**, 69 (1974)
30 A. I. Anselm, *Introduction to Theory of Semiconductors*, Izol. Fiz. Mat. Lit., Moscow (1962)
31 B. D. Fried and S. D. Conte, *The Plasma Dispersion Function*, Academic, New York (1961)
32 L. D. Landau and E. M. Lifshitz, *Mechanics*, Pergamon, London (1976)
33 J. J. Hopfield, Phys. Rev. **112**, 1555 (1958)

Radiowaves and Polaritons in Anisotropic Media. Roland H. Tarkhanyan and Nikolaos K. Uzunoglu

ISBN: 3-527-40615-8

34 S. I. Pekar, *Crystal Optics and Additional Light Waves*, Naukova Dumka, Kiev (1982)

35 B. Tell and R. Martin, Phys. Rev. **167**, 381 (1968)

36 M. Hashimoto and I. Akasaki, Phys. Lett. **25A**, 38 (1967)

37 C. Olson and D. Linch, Phys. Rev. **177**, 1231 (1969)

38 T. McMahon and R. Bell, Phys. Rev. **182**, 526 (1969)

39 N. V. Bozhovskaia and V. K. Subashiev, Sov. Phys. Solid State **14**, 1875 (1972)

40 V. V. Karmazin and V. K. Miloslavski, Sov. Phys. Opt. **27**, 78 (1969)

41 L. E. Gurevich and R. H. Tarkhanyan, Sov. Phys. Semicond. **6**, 605 (1972)

42 L. D. Landau and E. M. Lifshitz, *Quantum Mechanics*, Pergamon, London (1977)

43 M. Ya. Azbel and E. A. Kaner, JETP **30**, 811 (1956)

44 V. L. Ginzburg and A. A. Rukhadze, *Waves in Magnetoactive Plasma*, Nauka, Moscow (1970)

45 A. Otto, Z. Phys. **216**, 398 (1968)

46 R. Ruppin and R. Englman, Rep. Prog. Phys. **33**, 149 (1970)

47 G. S. Agarval, Phys. Rev. B **8**, 4768 (1973)

48 V. Briksin, D. Mirlin and Yu. Firsov, Sov. Phys. Usp. **17**, 305 (1974)

49 G. Borstel, H. Falge and A. Otto, Springer Tracts Mod. Phys. **74**, 107 (1974)

50 V. M. Agranovich, Sov. Phys. Usp. **18**, 99 (1975)

51 R. Fuchs and K. L. Kliewer, Phys. Rev. A **140**, 2076 (1965)

52 K. L. Kliewer and R. Fuchs, Phys. Rev. **144**, 495 (1966)

53 R. Englman and R. Ruppin, J. Phys. C **1**, 614, 630, 1515 (1968)

54 R. Wallis and J. Brion, Solid State Commun. **9**, 2099 (1971)

55 K. W. Chiu and J. J. Quinn, Phys. Lett. A **35**, 469 (1971)

56 A. Hartstein, E. Burstein, J. Brion and R. Wallis, Surf. Sci. **34**, 81 (1973)

57 R. Wallis, J. Brion, E. Burstein and A. Hartstein, Phys. Rev. B **9**, 3424 (1974)

58 G. Borstel and H. Falge, Phys. Stat. Sol. (b) **83**, 11 (1977)

59 G. Puchkovskaya and V. Strizhevskii, Phys. Stat. Sol. (b) **89**, 27 (1978)

60 L. E. Gurevich and R. H. Tarkhanyan, Sov. Phys. Solid State **17**, 1273 (1975)

61 R. H. Tarkhanyan, Phys. Stat. Sol. (b) **72**, 111 (1975)

62 V. Bryxin, D. Mirlin and I. Reshina, Solid State Commun. **11**, 695 (1972)

63 W. Anderson, R. Alexander and R. Bell, Phys. Rev. Lett. **27**, 1057 (1971)

64 D. Evans, S. Ushioda and J. McMullen, Phys. Rev. Lett. **31**, 369 (1973)

65 D. L. Mills and A. A. Maradudin, Phys.
Rev Lett. **31**, 372 (1973)

66 V. V. Bryxin, Yu. M. Gerbshtein and D. Mirlin, Phys. Stat. Sol. (b) **51**, 901 (1972)

67 R. Reather, Phys. Thin Films **9**, 145 (1977)

68 R. H. Tarkhanyan, Sov. Phys. Solid State **32**, 1115 (1990)

69 R. Dragila, B. Luther-Davies and S. Vukovich, Phys. Rev. Lett. **55**, 1117 (1985)

70 R. R. Ramazashvili, JETP Lett. **43**, 298 (1986)

71 R. H. Tarkhanyan, Sov. Phys. Solid State **30**, 361 (1988)

72 V. M. Agranovich and D. L. Mills (eds.), *Surface Polaritons*, North-Holland, Amsterdam (1982)

73 R. E. Camley and D. L. Mills, Phys. Rev. B **26**, 1280 (1982)

74 E. Kretschmann, Z. Phys. **241**, 313 (1971)

75 K. M. Haussler, Phys. Stat. Sol. (b) **105**, K81 (1981)

76 R. H. Tarkhanyan, Int. J. Infrared Millim. Waves **15**, 739 (1994)

77 R. H. Tarkhanyan, Lith. J. Phys. **35**, 587 (1995)

78 R. H. Tarkhanyan and N. K. Uzunoglu, Prog. Electromagn. Res. (PIER) **29**, 321 (2000)

79 J. Wy, P. Hawrilak and G. Eliasson, Solid State Commun. **58**, 795 (1986)

80 A. Tselis and J. J. Quinn, Phys. Rev. B **29**, 2021 (1986)

81 V. M. Agranovich and V. E. Kravtsov, Solid State Commun. **55**, 85 (1985)

82 T. Ando, A. Fauler and F. Stern, Rev. Mod. Phys. **54**, 437 (1982)

83 K. Von Klitzing, G. Dorda and M. Pepper, Phys. Rev. Lett. **45**, 494 (1980)

84 R. E. Prange and S. M. Girvin (eds.), *The Quantum Hall Effect*, Springer, New York (1990)
85 R. H. Tarkhanyan, in Proc. 2nd National Conf. Semiconductor Microelectronics, Yerevan State University, 1999, pp. 87–91
86 J. B. Gunn, Solid State Commun. **1**, 88 (1963); IBM J. Res. Dev. **8**, 141 (1964)
87 V. L. Bonch-Bruevich, I. Zvyagin and A. Mironov, *Domain Electric Instability in Semiconductors*, Nauka, Moscow (1972)
88 B. K. Ridley, *Quantum Processes in Semiconductors*, Clarendon, Oxford (1982)
89 M. E. Levinshtein, Yu. K. Pozhela and M. S. Shur, *Gunn Effect*, Soviet Radio, Moscow (1975)
90 E. M. Conwell, *High Field Transport in Semiconductors*, Academic, New York (1967)
91 B. G. Streetman, *Solid State Electronic Devices*, Prentice-Hall, New York (1980)
92 K. Hess, *Advanced Theory of Semiconductor Devices*, Prentice-Hall International Editions, London (1988)
93 B. K. Ridley and T. B. Watkins, Proc. Phys. Soc. **78**, 2939 (1961)
94 C. Hilsum, Proc. IRE **50**, 185 (1962)
95 K. Hess, H. Morkoc, H. Shichijo and B. G. Streetman, Appl. Phys. Lett. **35**, 469 (1979)
96 A. C. Baynham, IBM J. Res. Dev. **13**, 568 (1969)
97 A. C. Baynham, J. Appl. Phys. **44**, 1247 (1973)
98 F. G. Bass, S. I. Khankina and V. M. Yakovenko, JETP **50**, 102 (1966)
99 L. E. Gurevich and I. V. Ioffe, Sov. Phys. Solid State **18**, 752 (1976)
100 R. H. Tarkhanyan, Sov. Phys. Solid State **18**, 2005 (1976)
101 R. H. Tarkhanyan, Solid State Commun. **21**, 745 (1977)
102 R. H. Tarkhanyan, Sov. Phys. Solid State **22**, 855 (1980)
103 R. H. Tarkhanyan, Sov. Phys. Solid State **22**, 1866 (1980)
104 R. H. Tarkhanyan and K. M. Karapetyan, Sov. Phys. Semicond. **17**, 742 (1983)
105 R. H. Tarkhanyan and K. M. Karapetyan, Sov. Phys. Surf. **7**, 61 (1988) [in Russian]
106 A. E. Stefanovich, Sov. Phys. Tech. Phys. **7**, 462 (1962)
107 A. B. Mikhailovskii, *Theory of Plasma Instabilities*, Plenum, London (1974)
108 Z. S. Gribnikov, Sov. Phys. Semicond. **6**, 1204 (1973)
109 M. Keever, H. Shichijo and K. Hess, Appl. Phys. Lett. **38**, 36 (1982)
110 J. Pozhela, *Plasma and Current Instabilities in Semiconductors*, Pergamon, Oxford (1981)
111 R. H. Tarkhanyan and K. M. Karapetyan, Sov. Phys. Tech. Phys. **37**, 175 (1992)
112 A. Karlsson and G. Kristensson, Radio Sci. **18**, 345 (1983)
113 P. E. Wannamaker, G. W. Hohmann and W. A. SanFilipo, Geophysics **49**, 60 (1984)
114 R. H. Hardman and L. C. Shen, Geophysics **51**, 800 (1986)
115 L. Tsang, E. Njoku and J. A. Kong, J. Appl. Phys. **46**, 5127 (1975)
116 W. P. Harokopus and P. B. Katehi, Int. J. Numer. Model. **4**, 3 (1991)
117 N. Fache, F. Olyslager and D. Zutter, *Electromagnetic and Circuit Modeling of Multiconductor Transmission Lines*, Pergamon, Oxford (1993)
118 G. Cavalcante, D. Rogers and A. Giarola, Radio Sci. **17**, 503 (1982)
119 A. K. Bhattacharyya, *Electromagnetic Fields in Multilayered Structures*, Artech House, Boston (1994)
120 V. G. Daniele and R. S. Zich, Radio Sci. **8**, 63 (1973)
121 G. K. Lee and J. A. Kong, Electromagnetics **3**, 111 (1983)
122 G. S. Bagby, D. P. Nyquist and B. C. Drachman, IEEE Trans. Microwave Theory Tech. **33**, 906 (1985)
123 D. H. S. Chang, Electromagnetics **6**, 171 (1986)
124 N. K. Das and D. M. Pozar, IEEE Trans. Microwave Theory Tech. **35**, 326 (1987)
125 L. Beyne and D. Zutter, ibid, **36**, 875 (1988)
126 E. W. Kolk, N. H. Baken and H. Blok, ibid, **38**, 78 (1990)
127 J. F. Kiang, S. M. Ali and J. A. Kong, ibid, **38**, 193 (1990)
128 S. Barkeshli and P. H. Pathak, ibid, **40**, 128 (1992)
129 S. G. Pan and I. Wolff, ibid, **42**, 2118 (1994)
130 P. Bernardi and R. Cicchetti, ibid, **42**, 1474 (1994)

131 K. A. Michalski and J. R. Mosig, IEEE Trans. Antennas Propag. **45**, 508 (1997)

132 F. V. Bunkin, JETP Lett. **32**, 338 (1957)

133 E. Arbel and L. B. Felsen, in *Electromagnetic Theory and Antennas*, ed. E. C. Jordan, Macmillan, New York (1963), pp. 391–420

134 Ch. P. Wu, IEEE Trans. Antennas Propag. **11**, 681 (1963)

135 J. L. Tsalamengas and N. K. Uzunoglu, J. Appl. Phys. **53**, 7149 (1982)

136 N. K. Uzunoglu, J. L. Tsalamengas and J. G. Fikioris, Radio Sci. **19**, 429 (1984)

137 J. R. Eshbach and R. W. Damon, Phys. Rev. **118**, 1208 (1960)

138 L. K. Brundle and N. J. Freedman, Electron. Lett. **4**, 132 (1968)

139 S. R. Seshadri, Proc. IEEE **58**, 506 (1970)

140 G. Cloet, Phys. Stat. Sol. **39**, K9 (1970)

141 T. Gerson and J. Nadan, IEEE Trans. Microwave Theory Tech. **22**, 918 (1974)

142 F. Bardati and P. Lampariello, ibid, **27**, 679 (1979)

143 D. Jones, *Theory of Electromagnetism*, Pergamon, Oxford (1964), pp. 60–63

144 R. E. Collin, *Field Theory of Guided Waves*, McGraw-Hill, New York (1960), Chap. 11

145 N. K. Uzunoglu, J. Opt. Soc. Am. **71**, 259 (1981)

Index

Radiowaves and Polaritons in Anisotropic Media. Roland H. Tarkhanyan and Nikolaos K. Uzunoglu

ISBN: 3-527-40615-8